Design Automation Tools and Software for Quantum Computing

Lukas Burgholzer · Robert Wille

Design Automation Tools and Software for Quantum Computing

Inside the Munich Quantum Toolkit

Lukas Burgholzer
Technical University of Munich and
Munich Quantum Software Company
Munich, Germany

Robert Wille
Technical University of Munich and
Munich Quantum Software Company
Munich, Germany

ISBN 978-3-032-06769-2   ISBN 978-3-032-06770-8   (eBook)
https://doi.org/10.1007/978-3-032-06770-8

This Springer imprint is published by the registered company Springer Nature Switzerland AG
The registered company address is: Gewerbestrasse 11, 6330 Cham, Switzerland

If disposing of this product, please recycle the paper.

Preface

Quantum computing, first envisioned by Richard P. Feynman in 1982, holds the promise of solving problems that lie beyond the capabilities of today's classical machines. It has the potential to revolutionize a wide range of fields in the twenty-first century, including cryptography, artificial intelligence, physics, and more. Once considered a distant dream, quantum computing has recently experienced remarkable progress—particularly in hardware development. Numerous companies and academic institutions are actively advancing quantum technologies, building systems with increasing numbers of qubits that are progressively more stable and less error-prone. As a result, quantum computers based on a variety of qubit technologies are now publicly accessible for experimentation, whether through supercomputing centers or cloud-based platforms.

However, the best hardware is only as effective as the software that enables meaningful applications—a lesson well learned from decades of research in designing and developing classical circuits and systems. Thanks to software tools and methods for *Electronic Design Automation* (EDA), it has become possible to create incredibly complex systems with billions of transistors that power our modern world. These automated design methods have been instrumental in making electronic systems ubiquitous in our daily lives by allowing efficient handling of complexity and performance optimization.

In contrast, most existing software solutions for quantum computing still heavily rely on manual approaches for designing and optimizing applications and preparing them for execution on actual quantum computers. This not only introduces a higher risk of errors, inefficiencies, and inconsistencies but also leaves decades of research on design automation for classical systems largely untapped.

The *Munich Quantum Toolkit* (MQT, [1]) is a comprehensive collection of software tools for quantum computing that explicitly leverages expertise in design automation. Its overarching goal is to provide robust solutions for design tasks across the entire quantum software stack. This includes high-level support for end users in realizing their applications, as well as efficient methods for the classical simulation, compilation, and verification of quantum circuits. Additionally, it offers tools for quantum error correction, support for physical design, and more. These capabilities are underpinned by advanced data structures (such as decision diagrams and the ZX-calculus) and core algorithmic techniques (such as SAT encodings and solvers). All tools developed as part of the MQT are available as open-source implementations and can be accessed at github.com/munich-quantum-toolkit.

This book offers a detailed look into some of the inner workings of this toolkit. For selected representative design tasks, we present innovative approaches that demonstrate how established and proven EDA concepts can be adapted to address the unique challenges of designing and realizing quantum applications and circuits. In particular, we focus on:

- classical simulation of quantum circuits,
- compilation of quantum circuits, and
- verification of quantum circuits.

Through a careful examination of these critical tasks, we show how automated design methods and corresponding software can significantly enhance the efficiency, scalability, and reliability of quantum computing. The solutions presented here highlight the substantial benefits of building upon established principles from classical circuit design, rather than reinventing the wheel with entirely new approaches. All tools and methods discussed in this book are freely available as open-source software within the MQT, providing a solid foundation for future advancements in quantum software development.

The underlying methods presented in this book are the result of several years of research conducted at the Chair for Design Automation at the Technical University of Munich and the Institute for Integrated Circuits at the Johannes Kepler University Linz. The toolkit is further supported by the Munich Quantum Software Company (MQSC) and is part of the Munich Quantum Software Stack (MQSS) ecosystem within the broader Munich Quantum Valley (MQV) initiative. We are deeply indebted to numerous individuals and institutions for their support—without which the MQT and this book would not have been possible.

In particular, we would like to thank Alwin Zulehner, Stefan Hillmich, Thomas Grurl, Hartwig Bauer, Sarah Schneider, Smaran Adarsh, Alexander Ploier, Tom Peham, Kevin Mato, Nils Quetschlich, Lucas Berent, Daniel Schönberger, Ludwig Schmid, Aaron Sander, Damian Rovara, Yannick Stade, Tobias Forster, Erik Weilandt, Laura S. Herzog, Patrick Hopf, Daniel Haag, Marcel Walter, and Simon Hofmann, as well as all other contributors to the MQT. Beyond those explicitly named, we extend our gratitude to all co-authors of the papers that laid the foundation for this book. We are also thankful to our home institutions and affiliated companies for fostering an encouraging and enjoyable research environment.

The development of the MQT has been supported by the European Research Council (ERC) under the European Union's Horizon 2020 research and innovation program (grant agreement No. 101001318), the Bavarian State Ministry for Science and Arts through the Distinguished Professorship Program, and the Munich Quantum Valley, which receives funding from the Bavarian state government through the Hightech Agenda Bayern Plus.

Finally, we would like to thank Springer Nature—and especially Charles "Chuck" Glaser—for making the publication of this work possible.

Munich, Germany	Lukas Burgholzer
March 2026	Robert Wille

Reference

1. R. Wille, L. Berent, T. Forster, et al., "The MQT handbook: A summary of design automation tools and software for quantum computing," *in IEEE International Conference on Quantum Software (QSW)*, 2024. https://doi.org/10.1109/QSW62656.2024.00013. arXiv: 2405.17543, A live version of this document is available at https://mqt.readthedocs.io

Contents

Introduction

Imagine a computer capable of solving problems beyond the reach of today's machines—a computer that exploits the fundamental laws of nature to manipulate information in ways that challenge our classical intuition. This is the promise of quantum computing [1]. At its core, quantum information theory provides the foundations for quantum computing, combining key concepts from quantum physics and computer science. While quantum physics describes the behavior of matter and energy at the smallest scales, computer science offers principles for systematically processing and transmitting information. The intersection of these fields—quantum information theory—yields new avenues for information processing and communication.

The concept of quantum computing was first articulated by Richard P. Feynman in 1982, who observed that classical computers struggle to efficiently simulate quantum systems [2]. Since then, significant progress has been made in understanding the theoretical principles, practical challenges, and enormous potential of quantum information processing. Today, quantum information theory and quantum computing are poised to transform fields such as cryptography, artificial intelligence, and fundamental physics.

Unlike classical computing, quantum computing is based on qubits—quantum bits—that leverage uniquely quantum phenomena. Qubits can exist in a superposition of logical states (0 and 1 simultaneously) and can display entanglement, where the state of one qubit is intrinsically linked to that of another, even across long distances. These features enable quantum computers to process information in ways that are fundamentally inaccessible to classical machines. However, qubits are inherently sensitive to errors from environmental noise and decoherence, which necessitates precise engineering and control in quantum hardware.

Recent years have seen rapid advances in quantum hardware, spearheaded by both industry and academia. Various physical implementations have been pursued to realize qubits, each capitalizing on different quantum phenomena such as charge, spin, polarization, or

L. Burgholzer and R. Wille, *Design Automation Tools and Software for Quantum Computing*, https://doi.org/10.1007/978-3-032-06770-8_1

frequency. Organizations such as IBM, IQM, Rigetti, and Google use superconducting circuits [3], while others like Xanadu or Quandela employ photonic qubits [4]. Alternative approaches include trapped ions [5] (IonQ, AQT, Quantinuum), neutral atoms [6] (QuEra, planqc, Pasqal), and silicon-based devices [7] (Intel). Increasingly, these companies are making their quantum hardware accessible via cloud platforms, enabling researchers and developers worldwide to implement and test quantum algorithms.

Yet, while significant resources are dedicated to advancing quantum hardware, realizing the potential of quantum computing equally depends on the availability of sophisticated software tools. Quantum software plays a crucial role in enabling end-users to design, implement, and test quantum programs and applications. Drawing a parallel to classical computing, quantum software must fulfill a similar role as electronic design automation (EDA) tools [8], which have been pivotal in the development of modern electronic systems by automating key aspects of the design process, such as synthesis, verification, optimization, and testing. EDA tools have greatly simplified the design of complex classical circuits, reducing development time and cost, and providing high-level abstractions for circuit design.

Similarly, quantum software must provide comprehensive support throughout the quantum design flow:

- Methods for algorithm design to discover efficient quantum solutions to computational problems.
- Synthesis techniques to construct complex quantum algorithms.
- Compilers that translate high-level quantum programs into hardware-specific instructions, accounting for constraints such as noise, connectivity, and gate fidelity.
- Error correction schemes that protect quantum states against environmental disturbances.
- Optimization procedures that minimize resource requirements and operational errors.
- Testing and verification strategies to ensure correctness and robustness of quantum programs.

Collectively, these tools are essential for developing robust and scalable quantum applications capable of making full use of quantum hardware.

Several software platforms have emerged to drive quantum computing research and development, including Qiskit [9], Cirq [10], the Microsoft Quantum Development Kit [11], or Pennylane [12]. These platforms offer a range of tools and interfaces to experiment with quantum algorithms and applications. However, many current quantum software solutions do not yet systematically integrate the design automation methodologies that are standard in classical circuit design. Instead, significant portions of the quantum design process remain manual, which can introduce inefficiencies and errors.

The *Munich Quantum Toolkit* (MQT, [13]) is a comprehensive collection of software tools for quantum computing that explicitly leverages expertise in design automation. Its overarching goal is to provide robust solutions for design tasks across the flow mentioned above. This includes high-level support for end users in realizing their applications, as well as efficient methods for the classical simulation, compilation, and verification of quantum circuits. Additionally, it offers tools for quantum error correction, support for physical design, and more. These capabilities are underpinned by advanced data structures (such as decision diagrams and the ZX-calculus) and core algorithmic techniques (such as SAT encodings and solvers). All tools developed as part of the MQT are available as open-source implementations and can be accessed at github.com/munich-quantum-toolkit.

This book offers a detailed look into some of the inner workings of this toolkit. For selected representative design tasks, we present innovative approaches that demonstrate how established and proven EDA concepts can be adapted to address the unique challenges of designing and realizing quantum applications and circuits. More precisely, three design tasks summarized in the following are covered.

Classical Simulation of Quantum Circuits

Quantum circuits are sequences of operations that are applied to an initial quantum state to perform quantum computations. The simulation of these circuits on classical machines is an essential part of the quantum computing research and development process, as it can, e.g., serve as a temporary substitute for actual quantum computers and provide a benchmark for measuring the advantages of quantum computers over conventional ones. However, this task is challenging because the quantum state and the operations are commonly represented by vectors and matrices that have an exponential size with respect to the number of qubits. Therefore, efficient methods and powerful resources are required to simulate quantum circuits classically. Part II of the book explores the use of decision diagrams, a dedicated data structure inspired by classical design automation, to accelerate simulation and overcome excessive memory requirements. In particular:

- Chapter 5 describes the first hybrid Schrödinger-Feynman quantum circuit simulation approach that works with decision diagrams. By using decision diagrams whenever they are most effective and resorting to arrays whenever they are not, the described scheme combines the best of both worlds and allows one to significantly advance the state of the art in decision diagram-based quantum circuit simulation.
- Chapter 6 describes the first framework to investigate the importance of the path chosen when simulating quantum circuits using decision diagrams, that is, the order in which individual computations are performed to arrive at the final result. While this is well understood for tensor networks—another commonly used data structure—it is hardly studied for decision diagrams. Instead of reinventing the wheel, a flow is established that allows one to reuse existing techniques from the domain of tensor networks. It is shown, both conceptually and experimentally, that choosing the right simulation path frequently makes the difference between a runtime on the order of hours and having the result available in the blink of an eye.

- Finally, Chap. 7 studies the connections between decision diagrams and tensor networks and provides a systematic analysis of their similarities and dissimilarities. These considerations lead to guidelines for choosing an appropriate data structure to classically simulate quantum circuits, depending on the respective use cases.

In general, the approaches described in Part II significantly improve the current state of the art in classical quantum circuit simulation and further establish decision diagrams as a core data structure for this task.

Compilation of Quantum Circuits

Similarly to the compilation of classical programs, any (high-level) quantum circuit needs to be *compiled* before being executed on an actual quantum computer. Quantum processors typically limit the types of gates which they can execute and the qubits to which these gates can be applied. Therefore, compilation must ensure that the resulting quantum circuit uses only gates that are supported by the device and only applies gates to qubits that are connected on the device.

Mapping quantum circuits to architectures that have limited connectivity between their qubits is a crucial step in the compilation flow, as it directly affects the feasibility and performance of the quantum circuit on a given device. It involves finding a way to map the qubits of a quantum circuit to the qubits of a quantum device, while minimizing the overhead of gates that are added to respect the limited connectivity of the device. This is an extremely challenging problem to solve due to its immense search space. Part III of the book explores design automation methods to determine *optimal* solutions to the quantum circuit mapping problem. The development of such solutions is crucial to establish lower bounds on the achievable performance and to evaluate the quality of the results from heuristic methods. More precisely:

- Chapter 10 describes the first *optimal* solution to the problem of mapping quantum circuits to architectures with limited connectivity between their qubits. The described technique uses a symbolic encoding and powerful reasoning engines to determine circuits that conform to the connectivity restrictions imposed by a device with minimal overhead. Due to the exponential complexity of the underlying problem, this approach is limited to circuits and architectures of rather small sizes. Nevertheless, a comparison to a heuristic algorithm provided by IBM's Qiskit shows that (at the time of publication) the circuits mapped by Qiskit required twice as many additional gates as necessary.
- Chapter 11 shows that the search space in optimal quantum circuit mapping can be drastically limited and that optimal results can still be obtained by using general observations that enable a substantial reduction of the number of permutations to be considered before each gate, which are the main cause of the huge complexity. These limitations are theo-

retically supported by group theory, and strategies are described to use them in existing methods for optimal mapping.

In general, the approaches described in Part III encode the respective problem in a symbolic fashion and use powerful reasoning engines to cope with the vast complexity of the underlying problem. Experiments using the described methods revealed that there is much room for improvement in existing heuristic solutions.

Verification of Quantum Circuits

The complex transformations involved in the compilation of quantum circuits substantially alter the structure of the underlying circuit throughout the design process. To ensure a reliable and error-free design flow, software solutions for efficiently and automatically checking the correctness of compilation results are becoming more and more important. This is similar in the classical realm, where efficient verification (more precisely, *equivalence checking*) methods are used to ensure the correctness of the design throughout the respective flow. But due to quantum effects such as superposition and entanglement, these methods cannot be used right out of the box to check for equivalence in the quantum realm. Part IV of the book investigates design automation methods for the verification of quantum circuits. In particular:

- Chapter 14 describes a general equivalence checking methodology that does not take the mentioned quantum effects (such as superposition and entanglement) as a burden but instead tries to explicitly take advantage of them to efficiently determine whether two quantum circuits are equivalent or not. Based on two complementary ideas (exploiting the reversibility of the considered quantum circuits and utilizing simulation), an equivalence checking flow is described that allows for far faster equivalency checking than ever before, consistently outperforming the previous state of the art.
- Chapter 15 builds on this methodology and tailors it towards the verification of compilation flow results. The resulting method represents the first scalable solution for ensuring the correctness of the quantum circuit compilation flow. It is demonstrated to verify instances with thousands of gates in seconds, even if optimizations are employed that are not directly accounted for. In contrast to the classical realm, this could eventually make verifying the results of sophisticated design flows feasible.
- Chapter 16 for the first time demonstrates how different stimuli generation schemes for simulative verification allow for a trade-off between expressiveness and runtime. It also demonstrates, theoretically and empirically, that many errors in quantum circuits can already be detected by a few simulations with randomly chosen initial states.
- Chapter 17 describes a complementary equivalence-checking approach based on the ZX-Calculus as a core data structure. Evaluations confirm that the described method based on

the ZX-Calculus is best used in tandem with the general methodology described above, as both approaches complement each other in a multitude of ways.

- Chapter 18 demonstrates how to handle a broader class of quantum circuits without reinventing the wheel. It describes the first methods for verifying the equivalence of so-called dynamic quantum circuits—circuits that use classical feedback during the execution of the circuit to influence the computation.
- Finally, Chap. 19 combines all these approaches to make the verification of parameterized quantum circuits possible, which form the basis for almost all near-term applications of quantum computing. This is achieved by combining a ZX-calculus approach working directly on parameterized circuits with an instantiation strategy to create parameter-free circuits that can be efficiently checked by existing equivalence checking methods. This presents the first solution for verifying this important class of quantum circuits.

In general, the approaches described in Part IV form the first comprehensive suite of efficient methods and automated tools for the verification of quantum circuits. By this, an important step towards avoiding or substantially mitigating the emerge of a verification gap for quantum circuits is taken, i.e., a situation where the physical development of a technology substantially outperforms our ability to design suitable applications for it or to verify it.

In order to provide proper context and keep this book self-contained, Chap. 2 offers the necessary background on quantum computing, and Chap. 3 reviews the core data structures and design automation methods that form the foundation of many tools presented throughout this book.

With this, the book provides a representative insight into the *Munich Quantum Toolkit*. It builds upon the original research works cited in the respective sections and chapters, specifically:

- **Classical Simulation of Quantum Circuits**: [14–16]
- **Compilation of Quantum Circuits**: [17–19]
- **Verification of Quantum Circuits**: [20–29].

Finally, as mentioned above, all tools and software introduced in this book are available as open-source packages on GitHub (https://github.com/munich-quantum-toolkit). This includes the following tools:

- **Classical Simulation of Quantum Circuits**: MQT DDSIM
  A quantum circuit simulator based on decision diagrams
  https://github.com/munich-quantum-toolkit/ddsim

- **Compilation of Quantum Circuits:** MQT QMAP
  A quantum circuit mapping tool powered by reasoning engines and heuristic search algorithms
  https://github.com/munich-quantum-toolkit/qmap

- **Verification of Quantum Circuits:** MQT QCEC
  A quantum circuit equivalence checking tool based on decision diagrams and the ZX-calculus
  https://github.com/munich-quantum-toolkit/qcec

These tools are powered by a state-of-the-art decision diagram package for quantum computing and a high-performance ZX-calculus library. Together with a comprehensive intermediate representation library for quantum computations these libraries form MQT Core [30]:

- **Data Structures and Core Methods:** MQT Core
  The backbone of the *Munich Quantum Toolkit (MQT)*
  https://github.com/munich-quantum-toolkit/core

All of the above tools have been mainly implemented in C++, but strive to be as user-friendly as possible for the community. Hence, push-button solutions are provided through Python bindings, pre-built Python wheels are available for all major platforms and Python versions, and all tools integrate natively with IBM's Qiskit. All tools are actively maintained and well documented.

References

1. M.A. Nielsen, I.L. Chuang, *Quantum Computation and Quantum Information* (Cambridge University Press, 2010)
2. R.P. Feynman, Simulating physics with computers. Int. J. Theor. Phys. **21**(6–7), 467–488 (1982). https://doi.org/10.1007/BF02650179
3. M.H. Devoret, A. Wallraff, J.M. Martinis, Superconducting qubits: a short review (2004). arXiv: cond-mat/0411174, Preprint
4. J.L. O'Brien, A. Furusawa, J. Vuković, Photonic quantum technologies. Nat. Photon **3**(12), 687–695 (2009). ISSN: 1749-4885, 1749-4893. https://doi.org/10.1038/nphoton.2009.229
5. F. Bernardini, A. Chakraborty, C. Ordóñez, Quantum computing with trapped ions: a beginner's guide (2023). [cond-mat, physics:physics, physics:quant-ph]. http://arxiv.org/abs/2303.16358, Preprint
6. K. Wintersperger, F. Dommert, T. Ehmer, et al., Neutral atom quantum computing hardware: performance and end-user perspective (2023). [quant-ph]. arXiv:2304.14360, Preprint
7. G. Burkard, T.D. Ladd, J.M. Nichol, A. Pan, J.R. Petta, Semiconductor spin qubits (2021). [cond-mat, physics:physics, physics:quant-ph]. arXiv: 2112.08863, Preprint
8. L.-T. Wang, Y.-W. Chang, K.-T. Cheng (eds.), *Electronic Design Automation* (Elsevier, 2009) ISBN: 978-0-12-374364-0. https://doi.org/10.1016/S1875-9661(08)X0006-4
9. A. Javadi-Abhari, M. Treinish, K. Krsulich, et al., *Quantum computing with Qiskit* (2024). https://doi.org/10.48550/arXiv.2405.08810. arXiv: 2405.08810 [quant-ph]
10. *Cirq: A python framework for creating, editing, and invoking Noisy Intermediate Scale Quantum (NISQ) circuits.* https://github.com/quantumlib/Cirq
11. *Quantum Development Kit*, Microsoft. https://microsoft.com/en-us/quantum/development-kit
12. V. Bergholm, J. Izaac, M. Schuld, et al., PennyLane: automatic differentiation of hybrid quantum-classical computations (2022). [physics, physics:quant-ph], arXiv: 1811.04968, Preprint
13. R. Wille, L. Berent, T. Forster, et al., The MQT handbook: a summary of design automation tools and software for quantum computing, in *IEEE International Conference on Quantum Software (QSW)* (2024). https://doi.org/10.1109/QSW62656.2024.00013. arXiv: 2405.17543, A live version of this document is available at https://mqt.readthedocs.io
14. L. Burgholzer, H. Bauer, R. Wille, Hybrid Schrödinger-Feynman simulation of quantum circuits with decision diagrams, in *Int'l Conference on Quantum Computing and Engineering* (2021). https://doi.org/10.1109/QCE52317.2021.00037
15. L. Burgholzer, A. Ploier, R. Wille, Exploiting arbitrary paths for the simulation of quantum circuits with decision diagrams, in *Design, Automation and Test in Europe* (2022)
16. L. Burgholzer, A. Ploier, R. Wille, Simulation paths for quantum circuit simulation with decision diagrams: what to learn from tensor networks, and what not. IEEE Trans. CAD Integr. Circuits Syst. (2022). https://doi.org/10.1109/TCAD.2022.3197969. arXiv: 2203.00703
17. R. Wille, L. Burgholzer, A. Zulehner, Mapping quantum circuits to IBM QX architectures using the minimal number of SWAP and H operations, in *Design Automation Conference* (2019). https://doi.org/10.1145/3316781.3317859
18. L. Burgholzer, S. Schneider, R. Wille, Limiting the search space in optimal quantum circuit mapping, in *Asia and South Pacific Design Automation Conference* (2022). https://doi.org/10.1109/ASP-DAC52403.2022.9712555
19. R. Wille, L. Burgholzer, MQT QMAP: efficient quantum circuit mapping, in *Int'l Symposium on Physical Design* (2023). https://doi.org/10.1145/3569052.3578928

20. L. Burgholzer, R. Wille, Improved DD-based equivalence checking of quantum circuits, in *Asia and South Pacific Design Automation Conference* (2020)
21. L. Burgholzer, R. Wille, The power of simulation for equivalence checking in quantum computing, in *Design Automation Conference* (2020)
22. L. Burgholzer, R. Wille, Advanced equivalence checking for quantum circuits. IEEE Trans. CAD Integr. Circuits Syst. (2021). https://doi.org/10.1109/TCAD.2020.3032630
23. L. Burgholzer, R. Raymond, R. Wille, Verifying results of the IBM Qiskit quantum circuit compilation flow, in *Int'l Conference on Quantum Computing and Engineering* (2020). https://doi.org/10.1109/QCE49297.2020.00051
24. L. Burgholzer, R. Kueng, R. Wille, Random stimuli generation for the verification of quantum circuits, in *Asia and South Pacific Design Automation Conference* (2021). https://doi.org/10.1145/3394885.3431590
25. T. Peham, L. Burgholzer, R. Wille, Equivalence checking paradigms in quantum circuit design: a case study, in *Design Automation Conference* (2022)
26. T. Peham, L. Burgholzer, R. Wille, Equivalence checking of quantum circuits with the ZX-Calculus. JETCAS (2022). https://doi.org/10.1109/JETCAS.2022.3202204
27. L. Burgholzer, R. Wille, Handling non-unitaries in quantum circuit equivalence checking, in *Design Automation Conference* (2022). https://doi.org/10.1145/3489517.3530482
28. T. Peham, L. Burgholzer, R. Wille, Equivalence checking of parameterized quantum circuits: verifying the compilation of variational quantum algorithms, in *Asia and South Pacific Design Automation Conference* (2023). https://doi.org/10.1145/3566097.3567932
29. L. Burgholzer, R. Wille, QCEC: a JKQ tool for quantum circuit equivalence checking. Softw. Impacts (2021)
30. L. Burgholzer, Y. Stade, T. Peham, R. Wille, MQT core: the backbone of the munich quantum toolkit (MQT). J. Open Source Softw. **10**(108), 7478 (2025). https://doi.org/10.21105/joss.07478

Quantum Computing

In order to keep this book self-contained, this chapter reviews the basics of quantum computing. It covers quantum states, quantum operations, and quantum circuits. Note that the respective basics are kept brief and only cover the aspects needed in all of the remaining chapters. Further topic-specific basics are provided later when needed. For a more detailed introduction to quantum computing itself, the reader is referred to [1].

2.1 Quantum States

In classical computing, the basic unit of information is called a *bit* that can assume either one of the two values 0 and 1. In quantum computing, the basic unit of information is called a quantum bit (*qubit*) and cannot only assume either one of two basis states, $|0\rangle$ and $|1\rangle$ (written in Dirac notation) but also any complex-valued linear combination (*superposition*) thereof, i.e., the state $|\psi\rangle$ of a qubit is defined by

$$|\psi\rangle = \alpha_0 |0\rangle + \alpha_1 |1\rangle, \text{ with } \alpha_0, \alpha_1 \in \mathbb{C}, \text{ and } |\alpha_0|^2 + |\alpha_1|^2 = 1. \qquad (2.1)$$

The α_i are called amplitudes and are frequently written in the form of a *state vector*

$$|\psi\rangle \equiv \begin{bmatrix} \alpha_0 \\ \alpha_1 \end{bmatrix}. \qquad (2.2)$$

While these amplitudes describe the behavior of a quantum system, they are no physical quantities that can be accessed. Instead, information from a qubit can only be extracted via *measurements* (i.e., observing the qubit). In contrast to classical computing, measuring a qubit causes its state to collapse to one of the basis states $|i\rangle$—each with probability $|\alpha_i|^2$.

© The Author(s), under exclusive license to Springer Nature Switzerland AG 2026
L. Burgholzer and R. Wille, *Design Automation Tools and Software for Quantum Computing*, https://doi.org/10.1007/978-3-032-06770-8_2

The basis for an n-qubit state is formed by the tensor product of single-qubit states, i.e.,

$$|i_{n-1}\rangle \otimes \cdots \otimes |i_0\rangle \equiv |i_{n-1} \ldots i_0\rangle \text{, with } i_j \in \{0, 1\} \text{ for } j \text{ from } 0 \text{ to } n - 1. \quad (2.3)$$

Consequently, the joint state of n qubits (also referred to as the system's *wave function*) can be described as a linear combination of these 2^n basis states, i.e.,

$$|\psi\rangle = \sum_{i=0}^{2^n-1} \alpha_i |i\rangle \text{ with } \alpha_i \in \mathbb{C} \text{ and } \sum_{i=0}^{2^n-1} |\alpha_i|^2 = 1, \quad (2.4)$$

This can again be interpreted as a state vector, i.e., $|\varphi\rangle \equiv [\alpha_0, \ldots, \alpha_{2^n-1}]^\top$. Measuring such a quantum state probabilistically collapses the system's state to one of the basis states— each with probability $|\alpha_i|^2$ for $i = 0, \ldots, 2^n - 1$.

Example 2.1 Consider a two-qubit quantum system whose state is described by the state vector

$$\tfrac{1}{\sqrt{2}}[1\,0\,0\,1]^\top \equiv \tfrac{1}{\sqrt{2}} |0\rangle + \tfrac{1}{\sqrt{2}} |3\rangle = \tfrac{1}{\sqrt{2}} |(00)_2\rangle + \tfrac{1}{\sqrt{2}} |(11)_2\rangle = \tfrac{1}{\sqrt{2}}(|00\rangle + |11\rangle). \quad (2.5)$$

This is a valid quantum state, since $|1/\sqrt{2}|^2 + |1/\sqrt{2}|^2 = 1$. Furthermore, it demonstrates a key phenomenon unique to quantum computing—*entanglement*. While the state of the complete system can be accurately described (by the above statevector), the state of the individual qubits cannot, i.e., the state $|\psi\rangle$ cannot be written as $|q_1\rangle \otimes |q_0\rangle$. Measurement of this state would yield $|00\rangle$ in half ($\alpha_{00} = |1/\sqrt{2}|^2 = 0.5$) of the cases, and $|11\rangle$ otherwise.

2.2   Quantum Operations and Gates

The state of any quantum system can be manipulated by quantum operations, also synony- mously referred to as *quantum gates*. Any quantum gate applied to the state of a quantum system must again yield a valid quantum state. Consequently, any such operation must be *unitary*, i.e., it must be a linear transformations $U : \mathbb{C}^{2^n} \to \mathbb{C}^{2^n}$ such that $U U^\dagger = I$, where $U^\dagger$ is the conjugate transpose of U and I is the identity transformation. In contrast to classical computing, this implies that any quantum operation U manipulating the state of a quantum system is inherently reversible—the inverse is given by the conjugate transpose $U^\dagger$. Typ- ically, these operations only operate on a small subset of a system's qubits. An operation acting on $k \leq n$ qubits (most frequently $k = 1$ or $k = 2$) is described by a $2^k \times 2^k$ unitary matrix U.

Example 2.2 Popular single-qubit gates include the Pauli gates X, Y, and Z, the Hadamard gate H, as well as the phase gate S. The respective matrices are

$$X = \begin{bmatrix} 0 & 1 \\ 1 & 0 \end{bmatrix} \quad Y = \begin{bmatrix} 0 & -i \\ i & 0 \end{bmatrix} \quad Z = \begin{bmatrix} 1 & 0 \\ 0 & -1 \end{bmatrix} \quad H = \frac{1}{\sqrt{2}} \begin{bmatrix} 1 & 1 \\ 1 & -1 \end{bmatrix} \quad S = \begin{bmatrix} 1 & 0 \\ 0 & i \end{bmatrix}. \quad (2.6)$$

Most multi-qubit gates are *controlled* gates, where a certain single-qubit gate is applied to a specified *target* qubit only if all designated *control* qubits are $|1\rangle$. A prominent example is the two-qubit controlled-NOT (*CNOT*), which is described by the matrix

$$CNOT(q_c, q_t) = \begin{bmatrix} 1 & 0 & 0 & 0 \\ 0 & 1 & 0 & 0 \\ 0 & 0 & 0 & 1 \\ 0 & 0 & 1 & 0 \end{bmatrix}, \quad (2.7)$$

and applies an X gate to the target qubit q_t whenever the control qubit q_c is in the $|1\rangle$ state.

It can be easily verified by matrix multiplication that all the gates mentioned above represent unitary transformations.

The action of a quantum operation U on a quantum state $|\psi\rangle$ corresponds to the matrix-vector product of the respective matrix with the vector representing the state,[1] i.e., $|\psi'\rangle = U |\psi\rangle$.

Example 2.3 The Hadamard gate maps the computational basis states $|0\rangle$ and $|1\rangle$ (often referred to as the *Z-basis*) in the following fashion:

$$H |0\rangle = \frac{1}{\sqrt{2}} |0\rangle + \frac{1}{\sqrt{2}} |1\rangle =: |+\rangle \quad (2.8)$$

$$H |1\rangle = \frac{1}{\sqrt{2}} |0\rangle - \frac{1}{\sqrt{2}} |1\rangle =: |-\rangle. \quad (2.9)$$

The resulting states, $|+\rangle$ and $|-\rangle$, form the so-called *X-basis*. The Hadamard gate can be used to convert between the Z and the X basis.

2.3   Quantum Computations and Circuits

Quantum computations are just sequences of quantum operations applied to the state of a system. Such computations are predominantly described by *quantum circuits*, which, just like their classical counterparts, are made up of a sequence of quantum gates. More precisely, a quantum circuit G is described as a sequence of gates

$$g_0 \cdots g_{|G|-1}, \quad (2.10)$$

---

[1] Technically, the matrix first needs to be extended to the full system size (by forming appropriate tensor products with identity matrices) for the multiplication to be applicable.

where $|G|$ denotes the number of gates in the circuit. Each gate g_i represents a corresponding unitary matrix U_i that is subsequently applied during the execution of a quantum circuit. As a result, the functionality of a quantum circuit $G = g_0 \ldots g_{|G|-1}$ can be obtained as a unitary *system matrix* U itself by determining

$$U = U_{|G|-1} \cdots U_0. \tag{2.11}$$

Example 2.4 Consider the quantum circuit $G = g_0 g_1$ acting on two qubits (denoted q_0 and q_1) with $g_0 = H(q_1)$ (i.e., an H gate applied to q_1) and $g_1 = CNOT(q_1, q_0)$ (i.e., a $CNOT$ gate with control qubit q_1 and target qubit q_0). Then, the respective matrices U_0 and U_1, as well as the overall matrix of the system $U = U_1 \cdot U_0$ are given by

$$U_0 = H \otimes I = \frac{1}{\sqrt{2}} \begin{bmatrix} 1 & 0 & 1 & 0 \\ 0 & 1 & 0 & 1 \\ 1 & 0 & -1 & 0 \\ 0 & 1 & 0 & -1 \end{bmatrix} \quad U_1 = \begin{bmatrix} 1 & 0 & 0 & 0 \\ 0 & 1 & 0 & 0 \\ 0 & 0 & 0 & 1 \\ 0 & 0 & 1 & 0 \end{bmatrix} \quad U = \frac{1}{\sqrt{2}} \begin{bmatrix} 1 & 0 & 1 & 0 \\ 0 & 1 & 0 & 1 \\ 0 & 1 & 0 & -1 \\ 1 & 0 & -1 & 0 \end{bmatrix}. \tag{2.12}$$

Quantum circuits are usually visualized through *quantum circuit diagrams*, where horizontal wires indicate the system's individual qubits, while gates that are placed on these wires indicate the sequence of operations to apply—operating from left to right.

Example 2.5 The quantum circuit described in the previous example can be illustrated as follows:

$$\tag{2.13}$$

To this end, the control of the $CNOT$ gate is indicated by $\bullet$, while the target is indicated by $\oplus$.

Given an initial state $|\psi_{init}\rangle$ (typically assumed to be $|0\ldots0\rangle$), performing the computation described by the circuit G involves applying all operations of the circuit to the initial state, i.e., computing

$$\left|\psi_{final}\right\rangle = U_{|G|-1} \times \cdots \times U_0 \left|\psi_{init}\right\rangle = U \left|\psi_{init}\right\rangle. \tag{2.14}$$

If this task is performed on a classical computer, it is commonly referred to as *classical quantum circuit simulation* (cf. Part II).

Example 2.6 Simulating the quantum circuit shown in the previous examples in this fashion results in the following computation:

$$U_1 \times U_0 \times |00\rangle = \begin{bmatrix} 1 & 0 & 0 & 0 \\ 0 & 1 & 0 & 0 \\ 0 & 0 & 0 & 1 \\ 0 & 0 & 1 & 0 \end{bmatrix} \times \frac{1}{\sqrt{2}} \begin{bmatrix} 1 & 0 & 1 & 0 \\ 0 & 1 & 0 & 1 \\ 1 & 0 & -1 & 0 \\ 0 & 1 & 0 & -1 \end{bmatrix} \times \begin{bmatrix} 1 \\ 0 \\ 0 \\ 0 \end{bmatrix} = \frac{1}{\sqrt{2}} \begin{bmatrix} 1 \\ 0 \\ 0 \\ 1 \end{bmatrix}. \qquad (2.15)$$

This precisely yields the state previously considered in Example 2.1.

Reference

1. M.A. Nielsen, I.L. Chuang, *Quantum Computation and Quantum Information* (Cambridge University Press, 2010)

Increasingly large quantum circuits can be executed reliably on actual quantum computers as a result of an increase in the number of qubits with an increase in coherence time and faster operations with higher fidelity. As computing power increases, so does the demand for software solutions and development tools to help end users and programmers maximize that potential. Similarly to the design of classical circuits and systems, implementing quantum algorithms on actual devices requires a variety of complex design tasks, such as classical simulation (cf. Part II), compilation (cf. Part III), and verification (cf. Part IV). Each of these design tasks poses a computationally challenging issue, whether because of the exponential size of the underlying representations of quantum states and operations or the enormous number of degrees of freedom. This means that effective data structures and methods are the backbone of any reliable design approach or tool. It would be difficult to tackle the enormous complexity of the underlying problems without them. This chapter (based on [1]) discusses the data structures and methods used to address these challenges throughout the remainder of this book. In particular, the following sections will review decision diagrams (Sect. 3.1), tensor networks (Sect. 3.2), the ZX-calculus (Sect. 3.3), and SAT solving (Sect. 3.4).

3.1   Decision Diagrams

Decision diagrams were introduced in the 1980s as a data structure for the efficient representation and manipulation of Boolean functions [2]. This led to the emergence of a wide variety of decision diagrams, including BDDs, FBDDs, KFDDs, MTBDDs, and ZDDs (see, for example, [3–8]), which made them a crucial tool in the development of modern circuits and systems. Due to their previous success, decision diagrams have been proposed for application in the realm of quantum computing [9–15]. Particularly for design tasks

L. Burgholzer and R. Wille, *Design Automation Tools and Software for Quantum Computing*, https://doi.org/10.1007/978-3-032-06770-8_3

like *simulation* [9, 15–17], *synthesis* [18–21], and *verification* [11, 22, 23] of quantum circuits, they recently attracted great attention. In fact, decision diagrams form the foundation for a large part of this book's contributions towards classical quantum circuit simulation (cf. Part II) and verification (cf. Part IV). A dedicated decision-diagram software package for quantum computing has been developed in C++ and is publicly available as part of the MQT Core library ([24], https://github.com/munich-quantum-toolkit/core). It provides the de-facto reference implementation of a fully fledged decision-diagram package for quantum computing.

This section (based on [25]) reviews the corresponding concepts.

3.1.1 Representation of Quantum States

First, we review how quantum states are represented using decision diagrams. To this end, we consider the simple case of a single-qubit system. As discussed in Chap. 2, the state $|\Psi\rangle$ of such a system is described by two complex-valued, normalized amplitudes α_0 and α_1, i.e.,

$$|\Psi\rangle = \alpha_0 \, |0\rangle + \alpha_1 \, |1\rangle , \tag{3.1}$$

which is commonly represented as a statevector

$$|\Psi\rangle \equiv \begin{bmatrix} \alpha_0 \ \alpha_1 \end{bmatrix}^{\mathsf{T}} . \tag{3.2}$$

A rather simple observation and consequence of Eq. 3.1 is that this vector can be equally split into a contribution of the $|0\rangle$ state (α_0) and a contribution of the $|1\rangle$ state (α_1), i.e.,

$$\overbrace{\left(\underset{|0\rangle}{\begin{bmatrix} \alpha_0 \end{bmatrix}} \ \underset{|1\rangle}{\begin{bmatrix} \alpha_1 \end{bmatrix}} \right)^{\mathsf{T}}}^{|\Psi\rangle} . \tag{3.3}$$

This decomposition is the core of the decision-diagram formalism. The decision diagram representing $|\Psi\rangle$ has the structure

$$|\Psi\rangle \equiv \begin{bmatrix} \alpha_0 \ \alpha_1 \end{bmatrix}^{\mathsf{T}} \equiv \quad . \tag{3.4}$$

It consists of a single *node* with one *incoming edge* that represents the entry point in the decision diagram, as well as two *successors* that represent the split shown in Eq. 3.3 and end in a *terminal* node (the black box). The state's amplitudes are annotated at the respective edges. Edges without annotations correspond to an edge weight of 1.

Example 3.1 Consider the computational basis states $|0\rangle$ and $|1\rangle$. Then, the corresponding decision diagrams have the structures

$$|0\rangle \equiv \begin{bmatrix} 1 & 0 \end{bmatrix}^{\mathsf{T}} \equiv \qquad\qquad |1\rangle \equiv \begin{bmatrix} 0 & 1 \end{bmatrix}^{\mathsf{T}} \equiv \qquad\qquad . \tag{3.5}$$

In each of the cases, one of the successors ends in the terminal node, while the other ends in a *zero stub* (indicated by a black dot)—uncannily resembling the corresponding vector descriptions.

Building off the intuition of a single-qubit state, we can move to larger systems.

Example 3.2 Consider the following statevector of a three-qubit system:

$$|\Psi\rangle = \begin{bmatrix} \frac{1}{2\sqrt{2}} & \frac{1}{2\sqrt{2}} & \frac{1}{2} & 0 & \frac{1}{2\sqrt{2}} & \frac{1}{2\sqrt{2}} & \frac{1}{2} & 0 \end{bmatrix}^{\mathsf{T}} \tag{3.6}$$

Then, $|\Psi\rangle$ can be recursively split into equally sized parts similar to Eq. 3.3, i.e.,

$$\overbrace{\underbrace{\overbrace{\underbrace{|000\rangle}_{\frac{1}{2\sqrt{2}}}\ \underbrace{|001\rangle}_{\frac{1}{2\sqrt{2}}}}_{|00q_0\rangle}\ \underbrace{\overbrace{|010\rangle\ |011\rangle}}_{|01q_0\rangle}}^{|0q_1q_0\rangle}\ \underbrace{\overbrace{\underbrace{|100\rangle\ |101\rangle}_{|10q_0\rangle}\ \underbrace{|110\rangle\ |111\rangle}_{|11q_0\rangle}}}^{|1q_1q_0\rangle}}^{|q_2q_1q_0\rangle}$$

$$\begin{bmatrix} \frac{1}{2\sqrt{2}} & \frac{1}{2\sqrt{2}} & \frac{1}{2} & 0 & \frac{1}{2\sqrt{2}} & \frac{1}{2\sqrt{2}} & \frac{1}{2} & 0 \end{bmatrix}^{\mathsf{T}}, \tag{3.7}$$

where $q_2, q_1, q_0 \in \{0, 1\}$. This directly translates to the decision-diagram formalism:

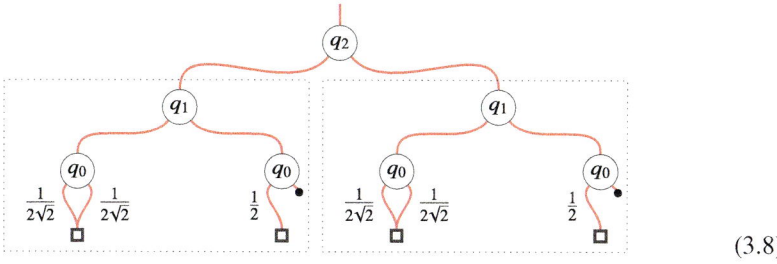

$$\tag{3.8}$$

Each level of the decision diagram consists of decision nodes with corresponding left and right successor edges. These successors represent the path that leads to an amplitude where the local quantum system (corresponding to the *level* of the node, annotated here with the labels) is in the $|0\rangle$ (left successor) or the $|1\rangle$ state (right successor).

At this point, this has been just a one-to-one translation between the statevector and a fancy graphical representation. The unique core feature of decision diagrams is that their graph structure allows redundant parts to be merged in the representation instead of being represented repeatedly.

Example 3.3 Observe how, as in the previous example, the left and right successors of the top-level node (labeled q_2) lead to exactly the same structure (highlighted by dashed rectangles in Eq. 3.8). As a result, the whole sub-diagram does not need to be represented twice, i.e.,

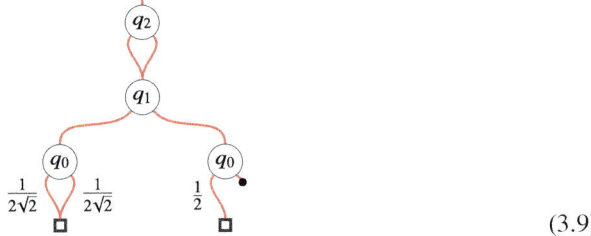

$$(3.9)$$

From a memory perspective, this reduction alone has compressed the overall memory required to represent the state by 50%.

Identifying redundancies in these kinds of representations heavily depends on the use of what is referred to as a *normalization scheme* for the decision-diagram nodes [12]. Such a normalization scheme makes sure that two decision-diagram nodes that represent the same functionality do indeed have the same numerical structure. In computer science, this property is called *canonicity* [12].

The most widely used and practically relevant normalization scheme is to normalize the outgoing edges of a node by dividing both weights by the norm of the vector containing both edge weights and adjusting the incoming edges accordingly [17]. This normalizes the sum of the squared magnitudes of the outgoing edge weights to 1 and is consistent with quantum semantics, where basis states $|0\rangle$ and $|1\rangle$ are observed after measurement with probabilities that are squared magnitudes of the respective weights. Normalization is recursively applied in a bottom-up fashion to ensure that every possible redundancy is caught.

Example 3.4 Considering the decision diagram from the previous example, this results in the following *normalized* and *reduced* decision diagram:

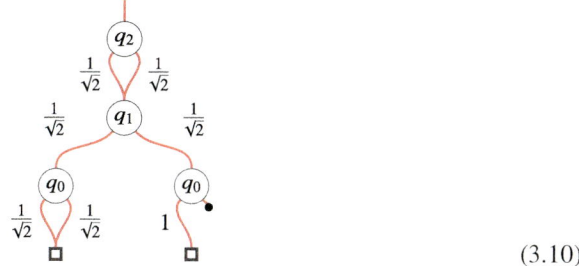

$$(3.10)$$

The first two levels (q_2 and q_1) of the above diagram naturally encode that the respective qubits have a 50/50 ($|1/\sqrt{2}|^2 = 0.5$) probability to be in $|0\rangle$ and $|1\rangle$. Meanwhile, the bottom level (q_0) encodes that the probability of q_0 depends on the state of q_1. If q_1 is in the $|0\rangle$ state (following the left successor), then q_0 has probability 0.5 in both $|0\rangle$ or $|1\rangle$. If q_1 is in the $|1\rangle$ state (following the right successor), it is guaranteed that the remaining qubit is in the $|0\rangle$ state.

Overall, statevectors are represented as decision diagrams conceptually equivalent to halving the vector in a recursive fashion until it is fully decomposed. The key idea is to exploit the redundancies in the resulting diagrams to create a more compact representation. Some interesting properties that are worth pointing out:

- Decision diagrams can be initialized in their compact form (as, e.g., shown in Example 3.4). There is no need to create the maximally large decision diagram (as shown, for example, in Example 3.2) at any point in a calculation.
- Determining a particular amplitude of the represented state corresponds to multiplying the edge weights along a single-path traversal from the top edge of the decision diagram (called its *root*) to a terminal node.
- The efficiency of decision diagrams is commonly measured by their *size*, i.e., the number of nodes in the decision diagram—the smaller the number of nodes, the higher the compaction achieved by the data structure. Note that the terminal (node) is typically not counted towards the size of a decision diagram.
- Any product state naturally has a decision diagram consisting of a single node per site. However, a compact DD does not correlate with the state being trivial. Even highly entangled states such as the GHZ state or the W state have decision diagrams whose size (i.e., the number of nodes) is linear in the number of qubits.
- DDs are not a "silver bullet." The worst-case size of decision diagrams, corresponding to states without redundancy, is still exponential in the number of qubits. More specifically, a maximally large decision diagram has $1 + 2^1 + 2^2 + \cdots + 2^{n-1} = 2^n - 1$ nodes.
- To reduce visual clutter in illustrations of decision diagrams, edge weights are commonly not explicitly annotated, but their magnitude and phase are reflected in the thickness and the color of the respective edge. In addition, to make the correspondence of the individual

levels in a decision diagram to a system's qubits more explicit, the nodes are frequently annotated with the qubit's index as an identifier. See [26] for further details on common techniques for visualization of decision diagrams.

3.1.2   Representing Quantum Operations

Quantum operations are fundamentally described by (complex-valued) matrices. Matrix decision diagrams are a natural extension to vector decision diagrams by an additional dimension. To this end, consider the base case of a 2×2 matrix U, i.e.,

$$U = \begin{bmatrix} U_{00} & U_{01} \\ U_{10} & U_{11} \end{bmatrix} = U_{00} \, |0\rangle\langle 0| + U_{01} \, |1\rangle\langle 0| + U_{10} \, |0\rangle\langle 1| + U_{11} \, |1\rangle\langle 1| . \tag{3.11}$$

Then, the decision diagram representing this matrix has the structure

$$U \equiv \qquad \tag{3.12}$$

which again resembles the general structure of the matrix. Note that U_{ij} can be interpreted as the transformation of $|j\rangle$ to $|i\rangle$.

Example 3.5 The following shows decision diagram representations for selected single-qubit operations:

$$I = \begin{bmatrix} 1 & 0 \\ 0 & 1 \end{bmatrix} \equiv \qquad \qquad X = \begin{bmatrix} 0 & 1 \\ 1 & 0 \end{bmatrix} \equiv$$

$$R_z(\theta) = \begin{bmatrix} e^{-i\frac{\theta}{2}} & 0 \\ 0 & e^{i\frac{\theta}{2}} \end{bmatrix} \equiv \qquad \equiv \tag{3.13}$$

The last equivalence demonstrates how a common factor between the edge weights can be pulled out and attached to the incoming (root) edge.

The generalization to larger matrices works analogously to the vector case. To construct the decision diagram representing a matrix, the matrix is recursively divided into quarters, and the four elements correspond to the four successors of the node to represent that split. As for vector decision diagrams, a normalization scheme is applied to ensure that the resulting data structure is canonical and redundancy can be exploited. The conventional approach is to normalize all edge weights by the weight with the highest magnitude, selecting the leftmost one if multiple weights have the same magnitude. It is important to note that this ensures that all complex numbers within the decision diagram have a magnitude of at most 1, although this is subject to implementation.

Example 3.6 Consider the maximally entangling two-qubit R_{xx} rotation represented by the matrix

$$R_{xx}\left(\theta = \frac{\pi}{2}\right) = \frac{1}{\sqrt{2}} \begin{bmatrix} 1 & 0 & 0 & -i \\ 0 & 1 & -i & 0 \\ 0 & -i & 1 & 0 \\ -i & 0 & 0 & 1 \end{bmatrix}. \tag{3.14}$$

This matrix is equivalent to blocks of 2×2 matrices corresponding to the identity I and the Pauli-X matrix, i.e.,

$$R_{xx}\left(\theta = \frac{\pi}{2}\right) = \frac{1}{\sqrt{2}} \begin{bmatrix} I & -iX \\ -iX & I \end{bmatrix}. \tag{3.15}$$

The corresponding (already reduced) decision diagram has the following structure:

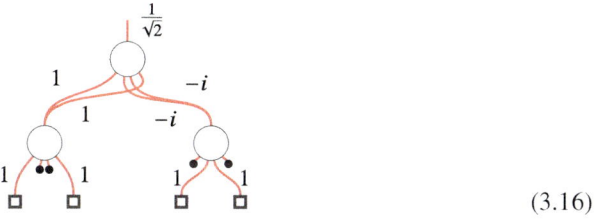

$$(3.16)$$

Notice how the decision diagram naturally resembles the structure of the matrix. The nodes at the bottom represent the identity and the X matrix (cf. Eq. 3.13) while the node at the top encodes the redundancy of the upper left quadrant and the bottom right quadrant, as well as the upper right and lower left quadrant in Eq. 3.15. Similarly to Example 3.3, exploiting redundancy has halved the overall memory requirement.

Again, some interesting properties to point out:

- Just as in the vector case, it is always possible to work with the reduced form of matrix decision diagrams right away, i.e., without ever constructing the exponentially sized, maximally large diagram.
- A maximally large matrix decision diagram for n qubits has $\sum_{i=1}^{n} 4^{i-1} = \frac{(4^n - 1)}{3}$ nodes.
- Decision diagrams are not limited to local interactions. Even long-range interactions between arbitrary qubits typically produce compact representations as decision diagrams.

For example, any two-qubit interaction between arbitrary qubits can be represented as a decision diagram with at most $1 + 4(n - 1)$ nodes—an exponential reduction.

• Decision diagrams are not limited to two-qubit interactions either. For example, controlled quantum gates with arbitrarily many controls (such as the multi-controlled Toffoli gate) give rise to decision diagrams with a linear number of nodes.

3.1.3   Fundamental Operations on Decision Diagrams

Merely defining means for compactly representing any kind of state or operation does not yet allow one to perform efficient computations. It is crucial to also define efficient means of working with or manipulating the resulting representations. In the following, it is demonstrated how the most fundamental operations can be carried out within the decision-diagram formalism and how they scale. The focus is mainly on how operations are realized on vectors, since the concepts extend from vectors to matrices in a straightforward fashion.

The main concept throughout all of these schemes is to recursively break the respective operations down into subcomputations. This decomposition then naturally matches the recursive decomposition of decision diagrams. As such, operations generally scale with the number of nodes in the involved decision diagrams.

Kronecker Product

The Kronecker product is necessary to create product states and to chain together local operations. For vectors, it can be expressed as

$$|\Psi\rangle \otimes |\Phi\rangle = \begin{bmatrix} \Psi_0 \, |\Phi\rangle \\ \Psi_1 \, |\Phi\rangle \end{bmatrix} = \begin{bmatrix} \Psi_0 \begin{bmatrix} \Phi_0 \\ \Phi_1 \end{bmatrix} \\ \Psi_1 \begin{bmatrix} \Phi_0 \\ \Phi_1 \end{bmatrix} \end{bmatrix}. \tag{3.17}$$

In the decision-diagram formalism, this is one of the simplest operations to perform and is done by simply replacing the terminal nodes of the first decision diagram with the root node of the second decision diagram. In case of the above example, this has the following form:

$$\tag{3.18}$$

As such, its complexity is linear in the number of nodes of the first decision diagram.

Addition

Standard vector addition can be recursively broken down according to

$$|\Psi\rangle + |\Phi\rangle = \begin{bmatrix} \Psi_0 \\ \Psi_1 \end{bmatrix} + \begin{bmatrix} \Phi_0 \\ \Phi_1 \end{bmatrix}$$

$$= w \begin{bmatrix} \alpha_0 \\ \alpha_1 \end{bmatrix} + w' \begin{bmatrix} \alpha_0' \\ \alpha_1' \end{bmatrix} = \begin{bmatrix} w\alpha_0 + w'\alpha_0' \\ w\alpha_1 + w'\alpha_1' \end{bmatrix}, \tag{3.19}$$

where w and w' are common factors of the terms in $|\Psi\rangle$ and $|\Phi\rangle$, respectively.

In the decision-diagram formalism, this corresponds to a simultaneous traversal of both decision diagrams from their roots to the terminal (multiplying edge weights along the way until the individual amplitudes are reached) and back again (accumulating the results of the recursive computations). More precisely,

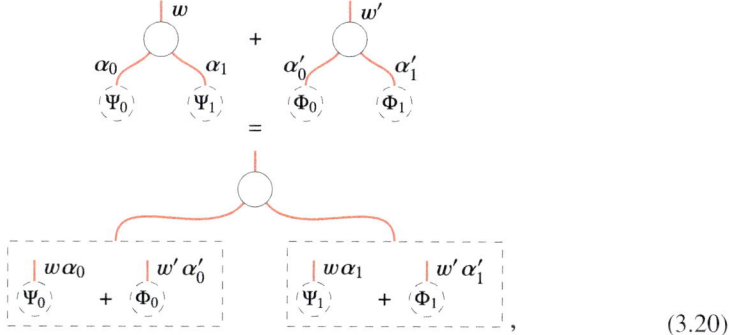

$$\tag{3.20}$$

where the dashed nodes represent the respective successor decision diagrams. Overall, this results in a complexity that is linear in the size of the larger decision diagram.

Matrix-Vector Multiplication

Matrix-vector multiplication can be handled in a very similar fashion as addition. Standard matrix-vector multiplication can be expressed as

$$U|\Psi\rangle = \begin{bmatrix} U_{00} & U_{01} \\ U_{10} & U_{11} \end{bmatrix} \begin{bmatrix} \Psi_0 \\ \Psi_1 \end{bmatrix}$$

$$= w \begin{bmatrix} u_{00} & u_{01} \\ u_{10} & u_{11} \end{bmatrix} w' \begin{bmatrix} \alpha_0 \\ \alpha_1 \end{bmatrix} = ww' \begin{bmatrix} u_{00} \cdot \alpha_0 + u_{10} \cdot \alpha_1 \\ u_{01} \cdot \alpha_0 + u_{11} \cdot \alpha_1 \end{bmatrix}. \tag{3.21}$$

This implies that a multiplication boils down to four smaller multiplications and two additions. In the decision-diagram formalism, this has the form

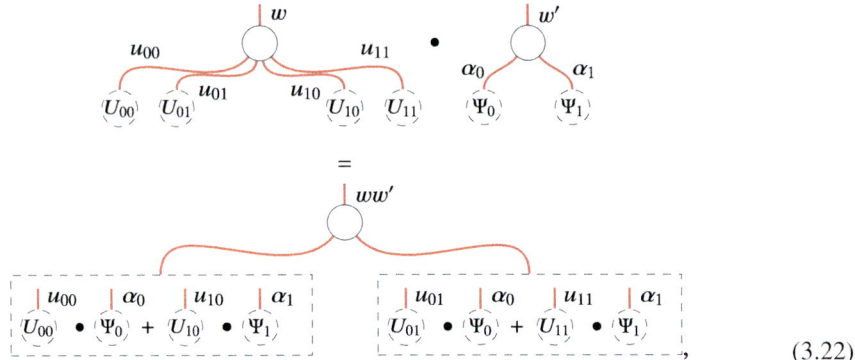

$$\begin{array}{ll}
\underbrace{\begin{array}{|c|c|c|c|}\hline u_{00} & \alpha_0 & u_{10} & \alpha_1 \\ \hline U_{00} & \Psi_0 & U_{10} & \Psi_1 \\ \hline\end{array}}_{U_{00}\ \bullet\ \Psi_0\ +\ U_{10}\ \bullet\ \Psi_1} &
\underbrace{\begin{array}{|c|c|c|c|}\hline u_{01} & \alpha_0 & u_{11} & \alpha_1 \\ \hline U_{01} & \Psi_0 & U_{11} & \Psi_1 \\ \hline\end{array}}_{U_{01}\ \bullet\ \Psi_0\ +\ U_{11}\ \bullet\ \Psi_1}
\end{array} \tag{3.22}$$

where the dashed nodes again represent the respective successor decision diagrams. Overall, this results in a complexity that scales with the product of the size of both decision diagrams.

Inner Product

Computing the inner product of two vectors can be recursively broken down according to

$$\langle \Psi | \Phi \rangle = \begin{bmatrix} \Psi_0^* & \Psi_1^* \end{bmatrix} \begin{bmatrix} \Phi_0 \\ \Phi_1 \end{bmatrix} \tag{3.23}$$

$$= w^* \begin{bmatrix} \alpha_0^* & \alpha_1^* \end{bmatrix} w' \begin{bmatrix} \alpha_0' \\ \alpha_1' \end{bmatrix} = w^* w' (\alpha_0^* \alpha_0' + \alpha_1^* \alpha_0')$$

This implies that the inner product boils down to two smaller inner product computations and adding the results. As with the matrix-vector multiplication, this is done recursively for each level of the decision diagram. In the decision-diagram formalism, this has the following form:

$$= w^* w' \left(\left\langle \Psi_0 \middle| \Phi_0 \right\rangle + \left\langle \Psi_1 \middle| \Phi_1 \right\rangle \right). \tag{3.24}$$

Overall, this results in a complexity that, just as in addition, scales linearly with the size of the larger decision diagram.

3.1.4 Visualizing Decision Diagrams

Decision diagrams are not yet widespread in the quantum computing community. As such, users of the corresponding tools often do not have an intuition about how methods based on decision diagrams work and what their strengths and limits are. In an effort to make decision diagrams for quantum computing more accessible, a tool has been developed which visualizes quantum decision diagrams and allows one to explore their behavior. The tool, called MQT DDVis, is publicly available at https://github.com/munich-quantum-toolkit/ddvis.

To make the tool as accessible as possible, it has been designed as a web tool that is hosted online at https://www.cda.cit.tum.de/app/ddvis/.

See [26] for further details.

3.2 Tensor Networks

In quantum many-body physics, which studies the collective behavior of interacting particles, particles close to another interact strongly, whereas particles at a distance hardly interact—inducing a notion of (topological) locality. This local structure naturally motivates the modeling of quantum many-body problems as *tensor networks* [27–29].

For the purpose of this book, a *tensor* can be understood as a multi-dimensional array of complex numbers. Here, the *rank* of a tensor is its number of dimensions (or indices), while the *shape* specifies the number of elements in each dimension. Two tensors that share common indices can be *contracted* into a single tensor by summing over repeated indices.

Example 3.7 Let A, B, C be matrices in $\mathbb{C}^{N \times N}$. Furthermore, let the matrix product $C = AB$ be given by $C_{i,j} = \sum_{k=0}^{N-1} A_{i,k} B_{k,j}$, with $i, j = 0, \ldots, N-1$. Then, this corresponds to the contraction of the rank-2 tensors $A = [A_{i,k}]$ and $B = [B_{k,j}]$ over the shared index k.

A *tensor network* is a countable set of tensors connected by shared indices. This is conveniently represented using a graphical notation, where individual tensors are represented as vertices of an undirected graph. The edges between the nodes describe the shared indices of the respective tensors.

Example 3.8 Consider again the situation as in Example 3.7. Then, the matrix multiplication $C = AB$ can be graphically represented as:

$$i \;—\!\boxed{C}\!—\; j \;=\; i \;—\!\boxed{A}\!\overset{k}{\rule{3em}{0.4pt}}\!\boxed{B}\!—\; j \tag{3.25}$$

On the basis of this, there is an intuitive translation between a quantum circuit and a tensor network. To this end, the initial state vector and the output state vector (if only individual amplitudes should be determined) are typically represented by sets of rank-1 tensors. In addition, each k-qubit gate is represented by a rank-$2k$ tensor that is connected to the tensors preceding/following it via shared indices. This is just a particular way to shape the underlying $2^k \times 2^k$-dimensional gate matrix that can be understood as having k ingoing and k outgoing edges, each of which has dimension 2.

Example 3.9 Consider the following quantum computation that aims to compute the α_{000} amplitude of a three-qubit GHZ state:

$$
\begin{array}{l}
q_2:\; |0\rangle \;—\!\boxed{H}\!—\!\bullet\!\rule{4em}{0.4pt}\; \langle 0| \\[4pt]
q_1:\; |0\rangle \;\rule{6em}{0.4pt}\!\oplus\!\rule{1em}{0.4pt}\!\bullet\!—\; \langle 0| \\[4pt]
q_0:\; |0\rangle \;\rule{9em}{0.4pt}\!\oplus\!—\; \langle 0|
\end{array}
\tag{3.26}
$$

Then, the corresponding tensor network has the following form:

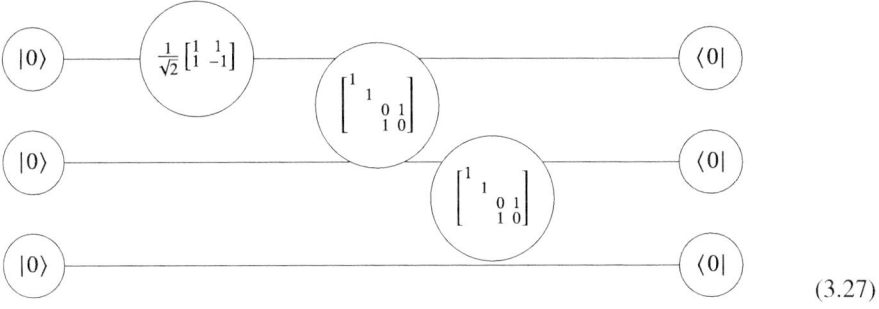

$$\tag{3.27}$$

To this end, each individual tensor is illustrated by a "bubble" containing the actual tensor data.

Extracting useful information from a tensor network typically requires pairwise contraction of all its individual tensors until only one tensor remains. In this regard, the goal is to choose an order of contractions that keeps the dimensions of contracted indices (also referred to as the *bond dimension*) and the shape of intermediate tensors as small as possible, because the computational effort required to contract two tensors is directly correlated with these quantities. Whenever the bond dimension and the size of intermediate tensors can be kept moderate, computations can be conducted efficiently using tensor networks. The order in which the tensors of a given network are contracted is called the *contraction plan*. The problem of determining an optimal order of contractions has been proven to be NP-hard [30]. Accordingly, trying to efficiently solve this challenging task for tensor networks has been a heavily researched topic for years [31–34].

Example 3.10 Contracting the tensor network shown in Example 3.9 eventually yields the desired amplitude $\alpha_{000} = \frac{1}{\sqrt{2}}$.

Similarly to decision diagrams, tensor networks efficiently represent the initial state of a quantum system as well as the individual operations in the form of a dedicated data structure. Then, they choose a certain order in which to contract the individual tensors. The performance of this contraction depends only on the size and shape of the individual tensors but not on the values (the data) in the tensors. On the one hand, this implies that an a-priori estimate of a particular contraction plan's performance can be efficiently inferred from the sizes and shaped of the tensors involved in all contractions. On the other hand, this also means that without a proper contraction plan, there is nothing to be gained by employing tensor networks. A systematic analysis on the (dis)similarities between tensor networks and decision diagrams will be provided later in Chap. 7.

3.3   ZX-Calculus

The ZX-calculus [35, 36] is a graphical notation for quantum circuits equipped with a powerful set of rewrite rules that enable diagrammatic reasoning about quantum computing. It has been successfully applied to the compilation and optimization of quantum circuits [37–39] and, as shown later in chapter 17, to verify the equivalence of quantum circuits.

A ZX-diagram is made up of colored nodes (called *spiders*) that are connected by wires (representing qubits, similar to quantum circuit notation). Each spider can either be green (Z-spider $\bigcirc$) or red (X-spider $\bullet$) and is optionally attributed a scalar phase.

ZX-diagrams can be composed just like quantum circuits. Horizontal composition or *concatenation* (denoted ∘) is achieved by connecting the outputs of one diagram to the input of another. The bare wire " ___ " acts as the identity for concatenation. Vertical composition (denoted ⊗) is achieved by simply "stacking" two diagrams on top of each other. The *empty diagram* " ⌐ ⌐ " acts as the identity for vertical composition. Additionally, a ZX-diagram can carry a global phase that is annotated along the diagram. Since global phases are negligible in most cases, they are frequently omitted from ZX diagrams, and equations in the ZX-calculus usually hold up to a global phase. A spider with a phase of $\pm\pi$ is called a *Pauli spider*. A spider with a phase $\alpha \in \{k\frac{\pi}{2} \mid k \in \mathbb{Z}\}$ is called a *Clifford spider*. A Clifford spider that is not a Pauli spider is called a *proper Clifford spider*. A ZX-diagram consisting entirely of Clifford (Pauli) spiders is called a *Clifford (Pauli) ZX-diagram*.

Any quantum circuit can be interpreted as a ZX-diagram. The reverse of this statement is not true, i.e., not every ZX-diagram can be interpreted as a quantum circuit because the ZX-diagram does not necessarily encode a unitary transformation. However, every ZX diagram has an interpretation as a linear map.

$$\llbracket\ \alpha\ \rrbracket = |0\ldots0\rangle\langle0\ldots0| + e^{i\alpha}|1\ldots1\rangle\langle1\ldots1| \tag{3.28}$$

$$\llbracket\ \alpha\ \rrbracket = |+\cdots+\rangle\langle+\cdots+| + e^{i\alpha}|-\cdots-\rangle\langle-\cdots-| \tag{3.29}$$

Here, $\llbracket\cdot\rrbracket$ denotes the *interpretation function* which maps a ZX-diagram to its corresponding linear map. Any linear map on qubits can then be built up from Z- and X-spiders by connecting and stacking diagrams. Spiders without inputs are called *states*, whereas spiders with no outputs are called *effects*. Interpreted as linear maps, states represent column vectors, whereas effects represent row vectors. A ZX-diagram without inputs or outputs represents a number.

The real power of ZX-diagrams becomes evident when adding rewrite rules to the language. The axioms of the scalar-free ZX-calculus are given in Fig. 3.1.

Example 3.11 To give a feel for how to work with ZX-diagrams, a derivation of the well-known equivalence of a SWAP gate with 3 CNOT operations is shown, i.e.,

$$\tag{3.30}$$

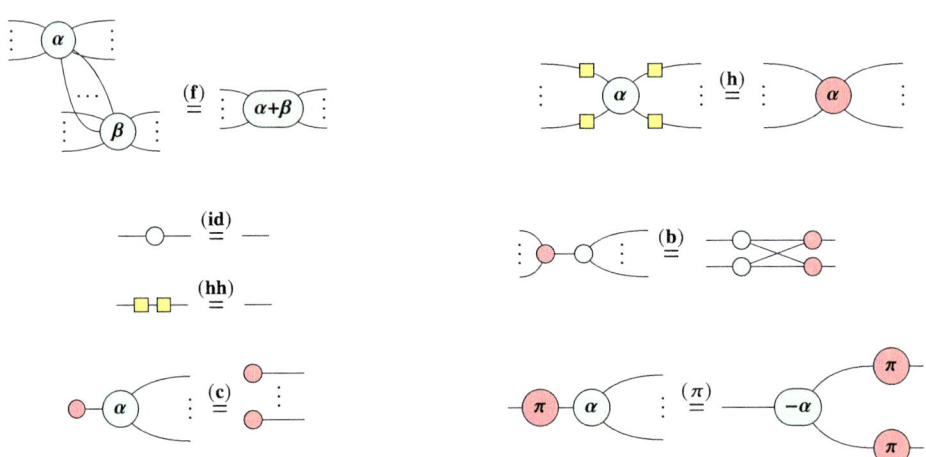

Fig. 3.1 Axioms of the scalar-free ZX-calculus

To this end, the following rule, which is sometimes listed explicitly among the axioms for the ZX-calculus, needs to be proven first:

$$\text{(3.31)}$$

This allows to proceed as follows:

$$\text{(3.32)}$$

As part of the Munich Quantum Toolkit (MQT, [40]), a dedicated ZX software package has been developed in C++ and is publicly available as part of the MQT Core library [24] at https://github.com/munich-quantum-toolkit/core. It can be seen as a high-performance version of the Python reference implementation PyZX (https://github.com/Quantomatic/pyzx) and an alternative to the C++-based ZX-calculus module that is part of Quantinuum's TKET (https://github.com/CQCL/tket).

3.4   SAT Solving

SAT (Satisfiability) solving [41] is the problem of determining an assignment of truth values to a given Boolean formula $\Phi : \{0, 1\}^n \rightarrow \{0, 1\}$ such that Φ evaluates to 1 or proving that no such assignment exists.

Example 3.12 Let $\Phi = (x_1 + x_2 + \overline{x}_3)(\overline{x}_1 + x_3)(\overline{x}_2 + x_3)$. Then, $x_1 = 1$, $x_2 = 0$, and $x_3 = 1$ is a satisfying assignment to solve the SAT problem.

As the first problem that was proven to be NP-complete [42], SAT solving is one of the most significant problems in theoretical computer science. This implies that any NP problem can be reduced to SAT in polynomial time. SAT solving also has numerous practical applications in which SAT solvers are utilized as high-performance reasoning engines for tasks such as automated planning and scheduling, formal verification, and automated theorem proving [43].

SAT solvers typically convert a given Boolean formula into *Conjunctive Normal Form* (CNF), which is a conjunction of clauses in which each clause is a disjunction of literals. For example, the formula Φ from Example 3.12 is already in the CNF. Then, SAT solvers use a variety of techniques and heuristics to either find a satisfactory assignment or prove none exists. Among the most prevalent algorithms is the DPLL algorithm [44, 45], which employs techniques such as unit propagation, pure literal elimination, and backtracking to perform a systematic search over the space of possible assignments.

Modern SAT solvers [43] include a variety of enhancements and features that increase their speed and scalability. These include heuristics for variable selection and value assignment, data structures that permit efficient implementation of search space pruning, preprocessing and inprocessing techniques that simplify the formula prior to and during the search, as well as learning mechanisms that record and reuse information regarding previous conflicts. In addition, parallel SAT solvers take advantage of the power of multiple processors or machines by executing multiple instances of a sequential SAT solver with different settings or solvers in parallel on the same problem, and exchanging information primarily in the form of learned clauses.

In addition to propositional logic, SAT solving issues can also be extended to incorporate richer logics and theories. This is accomplished by integrating SAT solvers with decision methods for specific theories such as linear algebra, bit vectors, arrays, or strings. The resulting solvers are known as Satisfiability Modulo Theories (SMT) solvers [41], and they can effectively reason about formulae that combine Boolean operators with predicates and functions from multiple theories. SMT solvers have a wide range of applications, including software verification, program synthesis, constraint resolution, and theorem proving.

MaxSAT solvers are extensions to standard SAT solvers that can tackle optimization problems that involve the maximization or minimization of a given objective function according to a set of constraints. In addition to *hard* constraints, which must be satisfied, MaxSAT formulations also include *soft* constraints, which need not necessarily be satisfied. The goal of MaxSAT solvers is to maximize the number of satisfied soft constraints, which can additionally be associated with a weight defining the relative importance of the constraints.

Example 3.13 Consider the formula Φ from Example 3.12 and, in addition, consider the objective function $\mathcal{F}$ defined by $\mathcal{F} := x_1 + x_2 + x_3$. Then, $x_1 = 0$, $x_2 = 0$, and $x_3 = 0$ is a solution which does not only satisfies Φ but also minimizes $\mathcal{F}$.

References

1. R. Wille, L. Burgholzer, S. Hillmich, T. Grurl, A. Ploier, T. Peham, The basis of design tools for quantum computing, in *Design Automation Conference* (2022)
2. Bryant, Graph-based algorithms for boolean function manipulation. IEEE Trans. Comput. **C-35**(8), 677–691 (1986)
3. R.E. Bryant, Symbolic boolean manipulation with ordered binary-decision diagrams. ACM Comput. Surv. **24**(3), 293–318 (1992)
4. I. Wegener, Branching programs and binary decision diagrams: theory and applications. SIAM J. Comp. (2000)
5. J. Gergov, C. Meinel, Efficient boolean manipulation with OBDD's can be extended to FBDD's. IEEE Trans. Comput. **43**(10), 1197–1209 (1994)
6. R. Drechsler, A. Sarabi, M. Theobald, B. Becker, M.A. Perkowski, Efficient representation and manipulation of switching functions based on ordered Kronecker functional decision diagrams, in *Design Automation Conference* (1994), pp. 415–419
7. R.I. Bahar, E.A. Frohm, C.M. Gaona, et al., Algebraic decision diagrams and their applications, in *Int'l Conference on CAD* (1993)
8. S. Minato, Zero-suppressed BDDs for set manipulation in combinatorial problems, in *Design Automation Conference* (1993) pp. 272–277
9. G.F. Viamontes, I.L. Markov, J.P. Hayes, Improving gate-level simulation of quantum circuits. Quantum Inf. Process. **2**(5), 347–380 (2003)
10. D. Miller, M. Thornton, QMDD: a decision diagram structure for reversible and quantum circuits, in *Int'l Symposium on Multi-Valued Logic* (2006)
11. S.-A. Wang, C.-Y. Lu, I.-M. Tsai, S.-Y. Kuo, An XQDD-based verification method for quantum circuits, in *IEICE Transactions on Fundamentals* (2008), pp. 584–594. https://doi.org/10.1093/ietfec/e91-a.2.584
12. P. Niemann, R. Wille, D.M. Miller, M.A. Thornton, R. Drechsler, QMDDs: efficient quantum function representation and manipulation. IEEE Trans. CAD Integr. Circuits Syst. (2016)
13. A. Zulehner, S. Hillmich, R. Wille, How to efficiently handle complex values? implementing decision diagrams for quantum computing, in *Int'l Conference on CAD* (2019)
14. X. Hong, X. Zhou, S. Li, Y. Feng, M. Ying, A tensor network based decision diagram for representation of quantum circuits (2020). arXiv: 2009.02618, Preprint
15. L. Vinkhuijzen, T. Coopmans, D. Elkouss, V. Dunjko, A. Laarman, LIMDD a decision diagram for simulation of quantum computing including stabilizer states (2021). arXiv: 2108.00931, Preprint
16. A. Zulehner, R. Wille, Advanced simulation of quantum computations. IEEE Trans. CAD Integr. Circuits Syst. (2019). https://doi.org/10.1109/TCAD.2018.2834427
17. S. Hillmich, I.L. Markov, R. Wille, Just like the real thing: fast weak simulation of quantum computation, in *Design Automation Conference* (2020)
18. P. Niemann, R. Wille, R. Drechsler, Efficient synthesis of quantum circuits implementing Clifford group operations, in *Asia and South Pacific Design Automation Conference* (2014), pp. 483–488
19. A. Abdollahi, M. Pedram, Analysis and synthesis of quantum circuits by using quantum decision diagrams, in *Design, Automation and Test in Europe* (2006)
20. M. Soeken, R. Wille, C. Hilken, N. Przigoda, R. Drechsler, Synthesis of reversible circuits with minimal lines for large functions, in *Asia and South Pacific Design Automation Conference* (2012), pp. 85–92
21. A. Zulehner, R. Wille, One-pass design of reversible circuits: combining embedding and synthesis for reversible logic. IEEE Trans. CAD Integr. Circuits Syst. **37**(5), 996–1008 (2018)

22. K.N. Smith, M.A. Thornton, Quantum logic synthesis with formal verification, in *IEEE International Midwest Symposium on Circuits and Systems* (2019). ISSN: 1548-3746. https://doi.org/10.1109/MWSCAS.2019.8885132

23. X. Hong, Y. Feng, S. Li, M. Ying, Equivalence checking of dynamic quantum circuits (2021). arXiv: 2106.01658, Preprint

24. L. Burgholzer, Y. Stade, T. Peham, R. Wille, MQT core: the backbone of the munich quantum toolkit (MQT). J. Open Source Softw. **10**(108), 7478 (2025). https://doi.org/10.21105/joss.07478

25. A. Sander, L. Burgholzer, R. Wille, Towards Hamiltonian simulation with decision diagrams, in *Int'l Conference on Quantum Computing and Engineering* (2023). https://doi.org/10.1109/QCE57702.2023.00039. arXiv: 2305.02337 [cond-mat, physics:quant-ph]

26. R. Wille, L. Burgholzer, M. Artner, Visualizing decision diagrams for quantum computing, in *Design, Automation and Test in Europe* (2021)

27. M. Fannes, B. Nachtergaele, R.F. Werner, Finitely correlated states on quantum spin chains. Commun. Math. Phys. **144**(3), 443–490 (1992)

28. J.D. Biamonte, V. Bergholm, Tensor networks in a nutshell (2017). arXiv: 1708.00006, Preprint

29. J.C. Bridgeman, C.T. Chubb, Hand-waving and interpretive dance: an introductory course on tensor networks. J. Phys. A: Math. Theor. (2017)

30. L. Chi-Chung, P. Sadayappan, R. Wenger, On optimizing a class of multi-dimensional loops with reduction for parallel execution. Parallel Process. Lett. (1997)

31. J. Gray, S. Kourtis, Hyper-optimized tensor network contraction. Quantum **5**, 410 (2021)

32. C. Huang, F. Zhang, M. Newman, et al., Classical simulation of quantum supremacy circuits (2020). arXiv: 2005.06787, Preprint

33. S. Boixo, S.V. Isakov, V.N. Smelyanskiy, H. Neven, Simulation of low-depth quantum circuits as complex undirected graphical models (2018). arXiv: 1712.05384, Preprint

34. D. Lykov, R. Schutski, A. Galda, V. Vinokur, Y. Alexeev, Tensor network quantum simulator with step-dependent parallelization (2020). arXiv: 2012.02430, Preprint

35. J. van de Wetering, ZX-calculus for the working quantum computer scientist (2020). arXiv: 2012.13966, Preprint

36. B. Coecke, A. Kissinger, Picturing quantum processes, in *Diagrammatic Representation and Inference*, ed. by P. Chapman, G. Stapleton, A. Moktefi, S. Perez-Kriz, F. Bellucci (2018)

37. A. Cowtan, W. Simmons, R. Duncan, A generic compilation strategy for the unitary coupled cluster ansatz (2020). arXiv: 2007.10515, Preprint

38. R. Duncan, A. Kissinger, S. Perdrix, J. van de Wetering, Graph-theoretic simplification of quantum circuits with the ZX-calculus (2019). arXiv: 1902.03178, Preprint

39. A. Kissinger, J. van de Wetering, Reducing T-count with the ZX-calculus. Phys. Rev. A (2020)

40. R. Wille, L. Berent, T. Forster, et al., The MQT handbook: a summary of design automation tools and software for quantum computing, in *IEEE International Conference on Quantum Software (QSW)* (2024). https://doi.org/10.1109/QSW62656.2024.00013. arXiv: 2405.17543, A live version of this document is available at https://mqt.readthedocs.io

41. A. Biere, M. Heule, H. van Maaren, T. Walsh, *Handbook of Satisfiability* (IOS Press, 2009)

42. S.A. Cook, The complexity of theorem-proving procedures, in *Symposium on Theory of Computing* (1971), pp. 151–158

43. S. Alouneh, S. Abed, M.H. Al Shayeji, R. Mesleh, A comprehensive study and analysis on SAT-solvers: advances, usages and achievements. Artif. Intell. Rev. **52**(4), 2575–2601 (2019). https://doi.org/10.1007/s10462-018-9628-0

44. M. Davis, H. Putnam, A computing procedure for quantification theory. J. ACM **7**(3), 201–215 (1960). https://doi.org/10.1145/321033.321034

45. M. Davis, G. Logemann, D. Loveland, A machine program for theoremproving. Commun. ACM **5**(7), 394–397 (1962). https://doi.org/10.1145/368273.368557

Classical Simulation of Quantum Circuits

Overview of Part II

Performing a quantum computation entails evolving an initial quantum state by applying a sequence of operations (also called gates) that is commonly described as a quantum circuit and measuring the resulting system. Eventually, the goal should obviously be to do that on a real machine. However, there are several important reasons for trying to simulate the corresponding computations on a classical machine, particularly in the early stages of the design: As long as no suitable machines are available (e.g., in terms of scale, feasible computation depth, or accuracy), classical simulation of quantum circuits still allows one to explore and test quantum applications, even if only on a limited scale. However, also with further progress in the capabilities of the hardware platforms, classical simulation will remain an essential part of the quantum computing design process, since it additionally allows access to *all* amplitudes of a resulting quantum state in contrast to a real machine that only probabilistically returns measurement results. Moreover, classical simulation provides means to study quantum error correction, as well as a baseline to estimate the advantage of quantum computers over classical computers.

Although conceptually simple (eventually, simulations can be conducted by subsequently multiplying matrices representing quantum operations with a vector representing a quantum state), simulating quantum circuits on classical machines is a formidable challenge. This is because the correspondingly considered vectors and matrices scale exponentially with respect to the number of qubits. For example, the full state of a 32-qubit system is described by $2^{32} = 4294967296$ complex amplitudes and storing it would require 64 GiB of memory (assuming a 128 bit representation for complex numbers). Consequently, quantum circuit simulation on classical machines in a straightforward fashion quickly becomes prohibitive for larger systems—even when resorting to powerful supercomputing clusters [1–5].

This part of the book explores the use of decision diagrams (cf. Sect. 3.1) as a dedicated data-structure to speed up the classical simulation of quantum circuits and overcome the

L. Burgholzer and R. Wille, *Design Automation Tools and Software for Quantum Computing*, https://doi.org/10.1007/978-3-032-06770-8_4

excessive memory requirements. In particular, it explores different simulation techniques, investigates how existing methods based on decision diagrams can be improved by exploiting knowledge and strategies from the domain of tensor networks (cf. Sect. 3.2), and studies the connection between these two data-structures. To this end, it gives an overview of several contributions towards various aspects of this domain in a cumulative fashion.

The precise contributions are as follows: First, a hybrid Schrödinger-Feynman simulation approach for decision diagrams is proposed (based on [6]) that allows one to split the simulation task into several simpler subtasks that can additionally be executed in parallel. Then, based on [7, 8], the importance of the order in which individual decision diagram computations are performed is studied—a topic well understood in the domain of tensor networks, but hardly considered for decision diagrams. Finally, a study (based on [9]) on the (dis)similarities between tensor networks and decision diagrams is presented that results in guidelines for when to better use tensor networks and when to better use decision diagrams in classical quantum circuit simulation.

References

1. T. Häner, D.S. Steiger, 0.5 petabyte simulation of a 45-Qubit quantum circuit, in *Int'l Conference for High Performance Computing, Networking, Storage and Analysis* (2017)
2. J. Doi, H. Takahashi, R. Raymond, T. Imamichi, H. Horii, Quantum computing simulator on a heterogenous HPC system, in *Int'l Conference on Computing Frontiers* (2019), pp. 85–93
3. T. Jones, A. Brown, I. Bush, S.C. Benjamin, QuEST and high performance simulation of quantum computers. Sci. Rep. (2018)
4. G.G. Guerreschi, J. Hogaboam, F. Baruffa, N.P.D. Sawaya, Intel quantum simulator: a cloud-ready high-performance simulator of quantum circuits. Quantum Sci. Technol. **5**(3), 034 007 (2020). ISSN: 2058-9565. https://doi.org/10.1088/2058-9565/ab8505
5. X.-C. Wu, S. Di, E. M. Dasgupta, et al., Full-state quantum circuit simulation by using data compression, in *Int'l Conference for High Performance Computing, Networking, Storage and Analysis* (2019)
6. L. Burgholzer, H. Bauer, R. Wille, Hybrid Schrödinger-Feynman simulation of quantum circuits with decision diagrams, in *Int'l Conference on Quantum Computing and Engineering* (2021). https://doi.org/10.1109/QCE52317.2021.00037
7. L. Burgholzer, A. Ploier, R. Wille, Exploiting arbitrary paths for the simulation of quantum circuits with decision diagrams, in *Design, Automation and Test in Europe* (2022)
8. L. Burgholzer, A. Ploier, R. Wille, Simulation paths for quantum circuit simulation with decision diagrams: what to learn from tensor networks, and what not, IEEE Trans. CAD Integr. Circuits Syst. (2022). https://doi.org/10.1109/TCAD.2022.3197969. arXiv: 2203.00703
9. L. Burgholzer, A. Ploier, R. Wille, Tensor networks or decision diagrams? guidelines for classical quantum circuit simulation (2023). arXiv: 2302.06616 [quant-ph], Preprint

Hybrid Schrödinger-Feynman Simulation of Quantum Circuits with Decision Diagrams

One of the most basic ways of simulating a quantum circuit is called *Schrödinger-style* simulation. This method stores and manipulates a complete representation of the state of the quantum system throughout the computation. However, this quickly becomes complex due to the underlying representation requiring the storage and manipulation of 2^n complex amplitudes for an n-qubit system. While this complexity is frequently tackled by massively parallel computations on arrays using supercomputer clusters with immense amounts of memory and processing power [1–5], decision diagrams (cf. Sect. 3.1) have been proposed as a complementary technique that aims at compactly representing and efficiently manipulating the 2^n complex amplitudes. This often allows corresponding simulations on a single desktop computer [6–11]. However, in the worst case, decision diagrams are still subject to the inherent exponential complexity.

Another way of simulating quantum circuits is *Feynman-style* path summation [12, 13]. This method reduces the memory complexity of the simulation by breaking it down into simpler parts. Each gate connecting two or more qubits in a quantum circuit introduces a decision point from which the simulation branches (this notion will become more precise later in Sect. 5.1.2). Feynman-style path summation calculates the result of each path and sums up all the individual contributions to obtain the final quantum state. Since the number of paths is exponential with regard to the number of decision points, this approach requires exponential run-time, but usually avoids too harsh memory requirements.

In recent years, *hybrid Schrödinger-Feynman* simulation has emerged as a mixture of both schemes [13–16]. These approaches use all available memory and processing units to efficiently simulate quantum circuits that would encounter memory bottlenecks using Schrödinger-style simulation or take an exceedingly long time using Feynman-style path summation. This eventually trades off the respective memory and runtime requirements. However, while this hybrid scheme can easily be realized using arrays, no solution for deci-

© The Author(s), under exclusive license to Springer Nature Switzerland AG 2026
L. Burgholzer and R. Wille, *Design Automation Tools and Software for Quantum Computing*, https://doi.org/10.1007/978-3-032-06770-8_5

sion diagrams exists yet. In fact, there are even discussions that a corresponding realization might not be possible at all [17]. This constitutes a severe drawback as it keeps the decision diagram-based simulation stuck with Schrödinger-style simulation, which is only suitable if the compact representations of decision diagrams allow one to escape the exponential memory requirements, while in all remaining cases, decision diagrams even impose a severe overhead compared to rather simple arrays.

In this chapter (based on [18]), it is shown that realizing a *hybrid Schrödinger-Feynman* scheme with decision diagrams is indeed possible—even if some problems arise when doing so. A possible realization is described, the problems that arise are discussed, and solutions to overcome them are described. Eventually, the first hybrid Schrödinger-Feynman quantum circuit simulation approach that works with decision diagrams will result. The experimental results demonstrate significant improvements compared to the state-of-the-art decision-diagram simulation approach. Circuits that could not be simulated in a whole day were simulated in approximately 20 min using the proposed method.

The remainder of this chapter is structured as follows: Sect. 5.1 describes the general idea of the hybrid Schrödinger-Feynman technique. Then, Sect. 5.2 describes the realization of such techniques for decision diagrams, the resulting limitations, and how they can be handled. Subsequently, Sect. 5.3 provides details on the implementation and a summary of the experimental results.

5.1 Motivation and General Idea

Decision diagrams offer a complementary approach to the simulation of quantum circuits. In many cases, they have been shown to compactly represent and efficiently manipulate the state of a quantum system. However, there are still some obstacles and limitations of decision-diagram-based simulation. In the following, these limitations will be discussed and a general idea for overcoming them will be provided.

5.1.1 Limitations of Decision Diagram-Based Simulation

While decision diagrams frequently allow one to compactly represent the state of a quantum system, in the worst case, their size is still exponential with respect to the number of qubits. Such situations arise when no redundancy in the description of the quantum state can be exploited and, thus, only a few nodes can be shared. This occurs, for example, during the simulation of quantum circuits whose gates are chosen randomly according to some scheme (e.g., the circuits used by Google in their quantum supremacy experiment [19, 20]). In general, such circuits are designed to make classical simulations as hard as possible, which—in case of decision diagrams—implies that they try hard to not give rise to a particular structure in the corresponding states.

At the same time, decision-diagram operations, such as matrix-vector multiplication, addition, inner product computation, or sampling, scale polynomially with the number of nodes of the involved decision diagrams (cf. Sect. 3.1). As such, they are highly efficient whenever the underlying decision diagrams actually emit a compact structure. However, if the number of nodes in the decision diagram grows exponentially (which happens in the worst case), their performance degrades significantly. Even worse: In this regime, decision diagrams actually perform worse than, e.g., array-based techniques, which always incur this exponential (memory) complexity, but have a lower overhead of maintaining the underlying data structure.

In one way or another, all Schrödinger-style methods (such as arrays, tensor trains, decision diagrams, etc.) face the problem of exponentially increasing simulation complexity. Many established simulation methods compensate for this limitation by heavily employing parallelization, that is, making use of the many cores in today's systems or even large clusters of supercomputers to speed up the computation [1–5]. Although similar efforts have been conducted to parallelize decision diagram-based simulation, e.g., in [21], no "breakthrough" has been achieved there yet. The main obstacles in this regard are that the shared nature of decision diagrams necessitates inter-process communication and some kind of locked access to its central data members (such as unique or compute tables), which, in turn, quickly eliminate the benefits of parallelization.

In this chapter, an approach is described that drastically reduces the exponential simulation complexity for certain classes of problems (specifically, depth-limited circuits) at the expense of simulating multiple, independent instances whose contributions are eventually accumulated. Coincidentally, this further allows fully utilizing all available processing power—effectively "killing both birds with one stone". The proposed approach follows the concepts of a *hybrid Schrödinger-Feynman technique*, which is reviewed next before the general idea of applying this concept to decision diagrams is described.

5.1.2   Hybrid Schrödinger-Feynman Simulation

The hybrid Schrödinger-Feynman simulation style aims at reducing the complexity of the Schrödinger-style simulation by breaking it down into simpler parts. This is accomplished by employing concepts from Feynman-style path summation. To this end, the most important concept is the Schmidt decomposition of a two-qubit gate: Any two-qubit gate (represented by a unitary 4×4 matrix U) can be decomposed into at most four tensor products according to

$$U = |0\rangle \langle 0| \otimes U_{00} + |0\rangle \langle 1| \otimes U_{01} + |1\rangle \langle 0| \otimes U_{10} + |1\rangle \langle 1| \otimes U_{11}, \qquad (5.1)$$

with unitary $U_{ij} \in \mathbb{C}^{2\times 2}$ and $i, j = 0, 1$.

Example 5.1 Consider the controlled-Z gate whose unitary matrix representation is given by $U = \mathrm{diag}(1, 1, 1, -1)$. Intuitively, the operation leaves both qubits untouched whenever the control qubit is in the $|0\rangle$ state, while applying a Z gate to the target qubit in the case when the control qubit is in the $|1\rangle$ state. In formulas:

$$|0\rangle \otimes |x\rangle \mapsto |0\rangle \otimes I\,|x\rangle \text{ and } |1\rangle \otimes |x\rangle \mapsto |1\rangle \otimes Z\,|x\rangle \text{ for } x = 0, 1. \tag{5.2}$$

Consequently, its Schmidt decomposition consists of two terms and is given by

$$U = |0\rangle\,\langle 0| \otimes I + |1\rangle\,\langle 1| \otimes Z = P_0 \otimes I + P_1 \otimes Z, \tag{5.3}$$

with P_0 (P_1) denoting the projection onto $|0\rangle$ ($|1\rangle$). This can be illustrated as:

$$\tag{5.4}$$

The Schmidt decomposition allows splitting the application of any two-qubit gate into separate parts that can be calculated independently. As such, each decomposed gate increases the number of simulations (i.e., the runtime) by the number of factors in its decomposition. After all individual contributions have been computed, they have to be summed up to obtain the full result.

Hybrid Schrödinger-Feynman approaches (such as [13–16]) horizontally partition the whole circuit into blocks by introducing *cuts* through the circuit's list of qubits. Only cross-block gates, i.e., gates acting across such a cut, are decomposed according to their Schmidt decomposition. In this way, individual blocks are independent of each other and, thus, can be simulated separately. Then, the total number of necessary simulations depends on the number of cross-block gates. As this dependence is exponential in the number of gates (e.g., doubling on each cross-block controlled-Z gate), such techniques are most efficient for depth-limited circuits. However, this still constitutes a large class of quantum algorithms—especially those targeted at near-term quantum computers, which are inherently depth limited due to noise.

5.1.3 General Idea

While decision diagrams offer a complementary approach to quantum circuit simulation that (exponentially) outperforms, for example, array-based techniques, whenever the number of nodes only grows polynomially, their performance is significantly worse for "hard" instances (where almost no redundancy can be exploited). The Schmidt decomposition, as introduced above, allows to reduce the complexity of individual simulations by splitting the circuit into independent blocks that are significantly easier to simulate at the expense of runtime. This concept can be readily applied to decision diagrams.

Example 5.2 Recall the controlled-Z gate and its decision diagram:

$$\boxed{\text{Z}} \quad = \quad \begin{bmatrix} 1 & & & \\ & 1 & & \\ & & 1 & \\ & & & -1 \end{bmatrix} \quad = \quad \text{} \tag{5.5}$$

As reviewed in Sect. 3.1, the successors of a (matrix) decision-diagram node encode its action according to the operator basis $\{P_0, |0\rangle\langle1|, |1\rangle\langle0|, P_1\}$. For the controlled-Z gate, the left-most successor (corresponding to P_0) leads to a node representing the identity operation while the right-most successor (corresponding to P_1) leads to a node representing the Z operation. Splitting these contributions into individual decision diagrams yields the decomposition:

$$\text{} \tag{5.6}$$

which precisely resembles the Schmidt decomposition of the controlled-Z gate from Example 5.1.

Overall, these techniques promise to overcome both obstacles raised above: By drastically reducing the complexity of individual simulations, the efficiency of decision diagrams can be fully exploited. Additionally, no inter-process communication or locked access is necessary when performing the simulations in parallel since they are independent of one another. Yet, hybrid Schrödinger-Feynman approaches have not been applied to decision diagrams. Some even believe that realizing such "circuit cutting techniques" with decision diagrams (as described in this chapter) is not possible at all [17].[1] In the remainder of this chapter it is shown that (1) this is indeed possible, (2) which problems arise in the realization, and (3) how they can be handled.

---

[1] In contrast to quantum circuit simulation as considered here, [17] seeks a complete representation of a quantum circuit's functionality. Both tasks are related as the functionality of a quantum circuit is obtained from consecutive matrix-matrix multiplication of the individual gate descriptions. Consequently, the results of this work are also applicable to the scenario discussed in [17].

5.2   Decision Diagram-Based Hybrid Schrödinger-Feynman Simulation

Following the general idea outlined above potentially allows one to overcome the limitations of decision diagram-based quantum circuit simulation discussed in Sect. 5.1.1. In this section, the realization of a hybrid Schrödinger-Feynman technique for decision diagrams is described, and the main bottleneck of the resulting scheme is discussed. Afterwards, it is shown how this bottleneck can be addressed by relying on decision diagrams where they are most efficient, while leaving the rest to more suitable techniques.

5.2.1   Realization

In order to employ a hybrid Schrödinger-Feynman approach, the circuit first has to be *partitioned* into blocks as reviewed in Sect. 5.1.2. In general, there is a great degree of freedom to choose such partitioning, e.g., the number of gates that act across individual blocks (*cross-block gates*). In the following, splitting the circuit into two (almost) equally sized blocks is considered—ensuring that each block approximately has the same number of qubits. Then, the number of simulations to be performed depends on the number of cross-block gates that need to be decomposed (according to their Schmidt decomposition). Each individual simulation can be assigned a unique identifier that specifies the *decision* (i.e., part of the Schmidt decomposition) to make at each *decision point* (i.e., cross-block gate).

Example 5.3 Consider the following circuit:

$$ \tag{5.7} $$

and assume it shall be partitioned into two equally sized halves. Then, both controlled-Z gates act across the blocks and, hence, need to be decomposed. This yields two decision points with two choices each (the parts of the controlled-Z gate's Schmidt decomposition)—for a total of four individual parts to be simulated as illustrated on the left-hand side of Fig. 5.1. To this end, the first (second) term of the Schmidt decomposition is encoded as 0 (1). Therefore, each simulation can be assigned a bitstring of length two (i.e., the number of decisions) that indicates which term of the decomposition is to be calculated.

Splitting the circuit in this way has three major benefits. Most importantly, the blocks in each individual simulation are independent from each other and thus can be simulated separately. This reduces the complexity for each simulation from simulating one n-qubit circuit to simulating two $n/2$-qubit circuits—an exponential improvement since the state of an n-qubit system grows as 2^n. Furthermore, as shown in Example 5.2, the decision diagrams for individual parts of a gate's Schmidt decomposition are typically much less complex than the full decision diagram—allowing for more compact decision diagrams throughout the simulations. Finally, all individual simulations are independent from each other and, hence, can be more efficiently simulated in parallel—even with decision diagrams.

Example 5.4 Consider again the scenario from Example 5.3. Then, each of the four individual simulations requires the simulation of two two-qubit circuits. The resulting decision diagrams are shown in the middle of Fig. 5.1. A maximum of two nodes (the best case for two-qubit decision diagrams) is required during each individual simulation.

Although this represents only a small example, it already shows the potential that the hybrid Schrödinger-Feynman technique brings to decision diagram-based simulation. However, the computation has not yet finished. As described in Sect. 5.1.2, the final result of the simulation is obtained as the sum of all individual parts. First, the simulation results of each

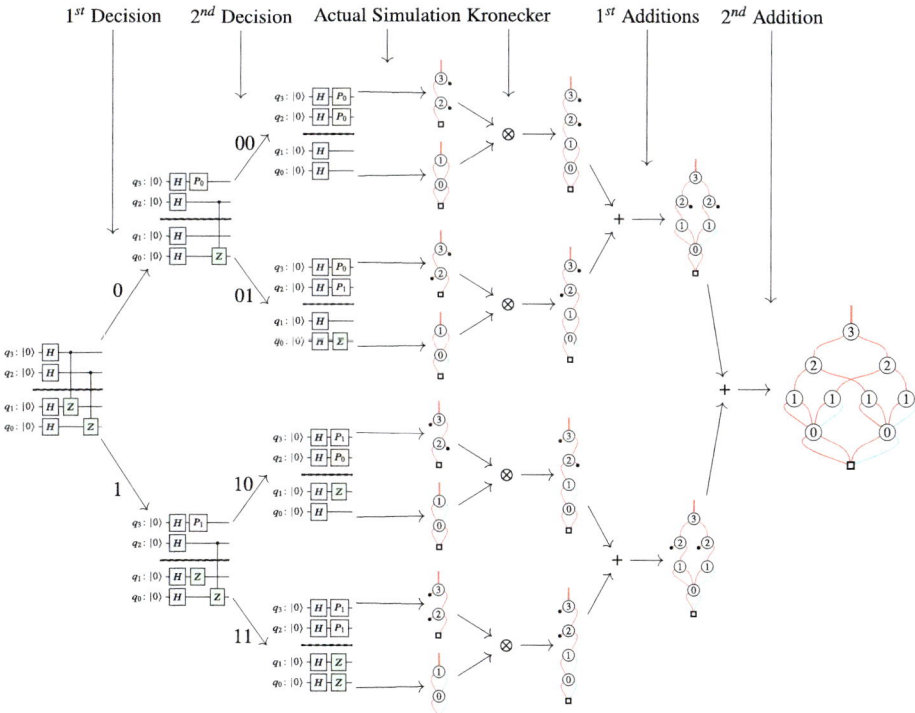

Fig. 5.1 Hybrid Schrödinger-Feynman simulation of the circuit shown in Example 5.3

block have to be combined by forming the tensor product of the corresponding states. The formation of the tensor product of two decision diagrams is highly efficient, as it merely requires replacing the terminal node of one decision diagram with the root node of the other.

Example 5.5 The formation of pairwise tensor products of the decision diagrams resulting from the circuit simulations from Example 5.4 yields four decision diagrams each of size four. These decision diagrams (shown in the middle of Fig. 5.1) represent the state vectors of the individual simulations.

Finally, all the decision diagrams need to be added up to obtain the decision diagram representing the final state vector. For the simulations, these additions can be computed in parallel without inter-process communication using a tree-like scheme whose depth corresponds to the logarithm of the number of individual simulations.

Example 5.6 The right-hand side of Fig. 5.1 illustrates the process of adding individual contributions to obtain the final state vector. At the first addition level, two decision diagrams of size six result, while after the final addition, the nine-node decision diagram that represents the final state results.

5.2.2   Decision-Diagram Addition as a Bottleneck

The combination of all individual results, i.e., the addition of all the decision diagrams that result, inevitably creates a potential bottleneck. While the complexity of the decision diagrams throughout the individual simulations might be drastically lower than the complexity of the complete Schrödinger-style simulation, the final decision diagram obviously remains the same. Consequently, whenever the final decision diagram grows exponentially, this complexity builds up somewhere along the way of adding up the individual contributions. Since the addition of the decision diagrams scales linearly with respect to the number of nodes in both decision diagrams, subsequent additions in the "addition hierarchy" take longer and longer.

At some point, this constitutes a severe bottleneck for the hybrid Schrödinger-Feynman simulation using decision diagrams because the overhead of maintaining the data structure becomes larger than the benefit gained from a potentially more compact representation. As confirmed by experimental evaluations (which are summarized later in Sect. 5.3), decision diagrams are highly efficient when it comes to the first part of the hybrid Schrödinger-Feynman scheme (i.e., the simulation of individual parts constituting the overall result), while it might get more challenging in the second part of the computation (i.e., adding up all the individual contributions). In the following, it is shown how this bottleneck can be addressed whenever the final decision diagram is going to be exponentially large.

5.2.3   Avoiding the Final Overhead

Thus far, decision diagrams have been used for the first part of the hybrid Schrödinger-Feynman scheme because of their efficiency in simulation. In the second part (when the final results are determined by addition), this benefit might disappear and lead to exponentially large decision diagrams. At this point, decision diagrams do not offer any advantages anymore compared to simpler data-structures such as arrays (in fact, the overhead caused by maintaining a dedicated data-structure will make decision diagrams perform even worse than arrays, which require practically no overhead). That is, in these cases, one is better off extracting the state vector represented by the decision diagram into an array and continuing working with that representation. That way, one relies on decision diagrams where they are most efficient while substituting more direct representations whenever the limit for decision diagrams has been reached.

To this end, the complete vector represented by a decision diagram is extracted with a single recursive traversal of the decision diagram by accumulating amplitude contributions along the edges. After the extraction, all resulting arrays can be added together to obtain the final state vector. As a consequence, one benefits from the memory locality of array-based representations as well as vectorized instruction support of modern CPUs—completely circumventing the overhead decision diagram-based addition incurs in this regime.

Example 5.7 Consider again the hybrid Schrödinger-Feynman scheme shown in Fig. 5.1 for simulating the circuit from Example 5.3. After all four individual simulations have been conducted, the amplitudes of the corresponding decision diagrams that represent the state vectors are extracted. For the top-most decision diagram (corresponding to the "00-decision") this results in the following (recursive) computation:

$$\frac{1}{2} * [[\cdots] \, 0000\,0000]^\top = \frac{1}{2} * [[\cdots] \, 0000\,0000\,0000]^\top$$
$$= \frac{1}{2} * [[\frac{1}{\sqrt{2}} * [\cdots] \frac{1}{\sqrt{2}} * [\cdots]] \, 0000\,0000\,0000]^\top$$
$$= \frac{1}{2\sqrt{2}} * [[\frac{1}{\sqrt{2}} \ \frac{1}{\sqrt{2}}][\frac{1}{\sqrt{2}} \ \frac{1}{\sqrt{2}}] \, 0000\,0000\,0000]^\top$$
$$= \frac{1}{4} * [1111\,0000\,0000\,0000]^\top \qquad (5.8)$$

Overall, the extraction results in the respective amplitude arrays

$$\frac{1}{4} * [\ 1\ \ 1\ \ 1\ \ 1\ \ 1\ \ 0\ \ 0\ \ 0\ \ 0\ \ 0\ \ 0\ \ 0\ \ 0\ \ 0\ \ 0\ \ 0\ \ ]^\top,$$
$$\frac{1}{4} * [\ 0\ \ 0\ \ 0\ \ 0\ \ 0\ \ 1{-}1\ \ 1{-}1\ \ 0\ \ 0\ \ 0\ \ 0\ \ 0\ \ 0\ \ 0\ \ 0\ \ ]^\top,$$
$$\frac{1}{4} * [\ 0\ \ 0\ \ 0\ \ 0\ \ 0\ \ 0\ \ 0\ \ 0\ \ 1\ \ 1{-}1{-}1\ \ 0\ \ 0\ \ 0\ \ 0\ \ ]^\top,$$
$$\frac{1}{4} * [\ 0\ \ 0\ \ 0\ \ 0\ \ 0\ \ 0\ \ 0\ \ 0\ \ 0\ \ 0\ \ 0\ \ 0\ \ 1{-}1{-}1\ \ 1\ \ ]^\top, \qquad (5.9)$$

which are subsequently added up to produce the final state vector

$$\tfrac{1}{4} * [\ 1\quad 1\quad 1\quad 1\quad 1-1\quad 1-1\quad 1\quad 1-1-1\quad 1-1-1\quad 1\ ]^{\top}. \qquad (5.10)$$

As experimental evaluations (which are summarized next) confirm, employing decision diagrams for the individual simulations of the hybrid Schrödinger-Feynman scheme, while handing off the accumulation of individual results to computations on arrays, allows to capitalize on the best of both worlds and mitigates the limitations discussed in Sect. 5.2.1.

5.3   Summary of Results

The decision diagram-based hybrid Schrödinger-Feynman simulation scheme described in Sect. 5.2 has been implemented on top of the Munich Quantum Toolkit's state-of-the-art decision diagram-based simulator MQT DDSIM. DDSIM is open source and publicly available at https://github.com/munich-quantum-toolkit/ddsim.

Its core is written in C++ and is based on the decision-diagram package, which has been co-developed as part of this book. To make the resulting methods as user-friendly as possible, the tool offers pre-built Python wheels for all major platforms and natively interfaces with IBM's Qiskit.

Extensive evaluations of the performance of the proposed simulation scheme have been conducted in [18], considering Google's supremacy circuits [20] as benchmarks. This led to the following results:

- The higher the number of qubits in the circuit, the higher the potential gain of employing the hybrid Schrödinger-Feynman scheme. Although using the general scheme described in Sect. 5.2.1 only yields a small speedup for smaller benchmarks (16 qubits), medium-sized benchmarks (20 qubits) show an average speedup of $\approx 6.2\times$.
- Neither the MQT DDSIM Schrödinger-style simulator, nor the general scheme described in Sect. 5.2.1 were able to simulate larger benchmarks (25 and 30 qubits) within 24 h. As discussed in Sect. 5.2.2, this can be attributed to the fact that decision-diagram addition on exponentially growing decision diagrams poses a severe bottleneck.
- This bottleneck is addressed by using decision diagrams for the individual simulations and resorting to arrays for the final additions. In fact, the results confirm that, then, speedups of several factors and up to several orders of magnitude can be observed across all benchmarks. More impressively, even the biggest circuits in the conducted evaluations that could not be simulated in a whole day using the Schrödinger-style simulator were simulated in roughly 20 min using this scheme.

Overall, these results showed that the proposed scheme combines the best of both worlds and significantly advances the state of the art in decision diagram-based quantum circuit simulation.

Implementation, Usage, Documentation, and Results

 The proposed methodology is available as part of the open-source MQT DDSIM tool at https://github.com/munich-quantum-toolkit/ddsim, which can be installed using pip install mqt.ddsim.

 Using the hybrid Schrödinger-Feynman simulation scheme in DDSIM to simulate a Qiskit quantum circuit is as easy as:

```python
from mqt.ddsim import DDSIMProvider
from qiskit import QuantumCircuit

qc = QuantumCircuit(4)
qc.h(range(4))
qc.cz(3, 1)
qc.cz(2, 0)

backend = DDSIMProvider().get_backend
("hybrid_qasm_simulator")
job = backend.run(circ, shots=1000,
mode="amplitude")
counts = job.result().get_counts(circ)
```

Here, mode = 'dd' | 'amplitude' and the nthreads argument can be used to control the number of parallel threads.

 Documentation on all available configuration options is available at https://mqt.readthedocs.io/projects/ddsim

 Details on the experimental setup, evaluations, and results can be found in [18].

References

1. T. Häner, D.S. Steiger, 0.5 petabyte simulation of a 45-Qubit quantum circuit, in *Int'l Conference for High Performance Computing, Networking, Storage and Analysis* (2017)
2. J. Doi, H. Takahashi, R. Raymond, T. Imamichi, H. Horii, Quantum computing simulator on a heterogenous HPC system, in *Int'l Conference on Computing Frontiers* (2019), pp. 85–93
3. T. Jones, A. Brown, I. Bush, S.C. Benjamin, QuEST and high performance simulation of quantum computers. Sci. Rep. (2018)
4. G.G. Guerreschi, J. Hogaboam, F. Baruffa, N.P.D. Sawaya, Intel quantum simulator: a cloud-ready high-performance simulator of quantum circuits. Quantum Sci. Technol. **5**(3), 034 007 (2020), ISSN: 2058-9565. https://doi.org/10.1088/2058-9565/ab8505
5. X.-C. Wu, S. Di, E.M. Dasgupta, et al., Full-state quantum circuit simulation by using data compression, in *Int'l Conference for High Performance Computing, Networking, Storage and Analysis* (2019)
6. A. Zulehner, R. Wille, Advanced simulation of quantum computations. IEEE Trans. CAD Integr. Circuits Syst. (2019). https://doi.org/10.1109/TCAD.2018.2834427
7. V. Samoladas, Improved BDD algorithms for the simulation of quantum circuits, in *Algorithms - ESA*, ed. by D. Halperin, K. Mehlhorn (2008)
8. G.F. Viamontes, I.L. Markov, J.P. Hayes, High-performance QuIDD-Based simulation of quantum circuits, in *Design, Automation and Test in Europe* (2004)
9. S. Hillmich, R. Kueng, I.L. Markov, R. Wille, As accurate as needed, as efficient as possible: approximations in DD-based quantum circuit simulation, in *Design, Automation and Test in Europe* (2020)
10. A. Zulehner, R. Wille, Matrix-vector versus matrix-matrix multiplication: potential in DD-based simulation of quantum computations, in *Design, Automation and Test in Europe* (2019). https://doi.org/10.23919/DATE.2019.8714836
11. T. Grurl, J. Fuß, S. Hillmich, L. Burgholzer, R. Wille, Arrays versus decision diagrams: a case study on quantum circuit simulators, in *Int'l Symposium on Multi-Valued Logic* (2020)
12. E. Bernstein, U. Vazirani, Quantum complexity theory. SIAM J. Comput. (1997)
13. S. Aaronson, L. Chen, Complexity-theoretic foundations of quantum supremacy experiments (2016). arXiv: 1612.05903, Preprint
14. Z.-Y. Chen, Q. Zhou, C. Xue, X. Yang, G.-C. Guo, G.-P. Guo, 64-qubit quantum circuit simulation. Sci. Bull. **63**(15), 964–971 (2018)
15. I.L. Markov, A. Fatima, S.V. Isakov, S. Boixo, Quantum supremacy is both closer and farther than it appears (2018). arXiv: 1807.10749, Preprint
16. E. Pednault, J.A. Gunnels, G. Nannicini, et al., Pareto-efficient quantum circuit simulation using tensor contraction deferral (2020). arXiv: 1710.05867, Preprint
17. X. Hong, X. Zhou, S. Li, Y. Feng, M. Ying, A tensor network based decision diagram for representation of quantum circuits (2020). arXiv: 2009.02618, Preprint
18. L. Burgholzer, H. Bauer, R. Wille, Hybrid Schrödinger-Feynman simulation of quantum circuits with decision diagrams, in *Int'l Conference on Quantum Computing and Engineering* (2021). https://doi.org/10.1109/QCE52317.2021.00037
19. S. Boixo, S.V. Isakov, V.N. Smelyanskiy, et al., Characterizing quantum supremacy in near-term devices. Nat. Phys. **14**(6), 595–600 (2018). arXiv: 1608.00263
20. S. Boixo, C. Neill, The question of quantum supremacy. Google AI Blog. (2018). https://github.com/sboixo/GRCS
21. S. Hillmich, A. Zulehner, R. Wille, Concurrency in DD-based quantum circuit simulation, in *Asia and South Pacific Design Automation Conference* (2020). https://doi.org/10.1109/ASP-DAC47756.2020.9045711

Sophisticated data structures such as tensor networks (cf. Sect. 3.2) or decision diagrams (cf. Sect. 3.1) have been demonstrated to alleviate the immense complexity of classical quantum circuit simulation in many practically relevant cases. In both cases, the efficiency is heavily dependent on the order in which the computations are performed. For tensor networks, this is known as the *contraction plan* of the tensor network and it is known that finding an optimal contraction plan is an NP-hard task [1]. For decision diagrams, the order of the individual matrix-matrix and matrix-vector multiplications is important, which will be referred to as the *simulation path* in the following. Although thoroughly investigated for tensor networks [2–5], the effect of the simulation path has hardly been studied for decision diagrams yet.

In this chapter (based on [6, 7]), this issue is investigated and a framework that allows one to exploit arbitrary simulation paths for decision diagram-based quantum circuit simulation is proposed. Instead of reinventing the wheel, a flow is established that not only allows to investigate dedicated paths but also to reuse existing techniques, e.g., from the tensor network domain, also for decision diagrams. Considering the verification of quantum circuit compilation flow results as a particularly important use case for quantum circuit simulation, it is shown, both conceptually and experimentally, that choosing the right simulation path can make a vast difference in the efficiency of classical simulations using decision diagrams. Based on numerical evaluations, the resulting consequences on what *can* be learned from tensor networks and what *cannot* be learned from them are discussed. This eventually provides the basis for future research on quantum circuit simulation using decision diagrams.

The remainder of this chapter is structured as follows: Sect. 6.1 illustrates the degrees of freedom and the potential impact of arbitrary simulation paths. Motivated by that, Sect. 6.2 presents the framework that allows one to evaluate these arbitrary simulation paths and describes how existing techniques from the tensor network domain can be used to obtain "good" simulation paths without starting from scratch. Subsequently, Sect. 6.3 summarizes the experimental results and provides a discussion of their implications.

6.1   Motivation and Related Work

This section first considers the question of how the order in which the respective multiplications involved in classically simulating a quantum circuit are conducted influences the complexity of simulation methods based on decision diagrams—a topic hardly considered thus far. Afterwards, an overview is given of how other types of simulators address this problem.

6.1.1   Considered Problem

Recall that the simulation of a quantum circuit $G = g_0 \ldots g_{|G|-1}$ given an initial state $|\varphi\rangle$ entails the sequence of computations

$$|\varphi\rangle G = |\varphi\rangle g_0 \ldots g_{|G|-1} \equiv U_{|G|-1} * \cdots * U_0 * |\varphi\rangle . \tag{6.1}$$

Performing these mutliplications using decision diagrams has been shown to allow efficient classical simulation of quantum circuits in many practical cases [8–13]. However, decision diagrams still exhibit exponential worst-case complexity.

Example 6.1 Let G be the circuit for the n-qubit quantum Fourier transform (QFT) that, given an initial state in the computational basis, outputs the state's representation in the Fourier basis. For $n = 3$ this circuit has the following form:

$$\tag{6.2}$$

In addition, let $|\varphi\rangle$ denote the n-qubit GHZ state defined by $\frac{1}{\sqrt{2}}(|0 \ldots 0\rangle + |1 \ldots 1\rangle)$. The corresponding decision diagram for $n = 3$ is given by

$$(6.3)$$

Assume that G shall be simulated with $|\varphi\rangle$ as input. Then, the representations of the decision diagram for the initial state $|\varphi\rangle$ as well as the individual gates are linear in the number of nodes. However, it can be shown that the decision diagram of the final state resulting from the simulation of G with initial state $|\varphi\rangle$ is maximally large, i.e., consists of $2^n - 1$ nodes. For $n = 3$ it has the following form:

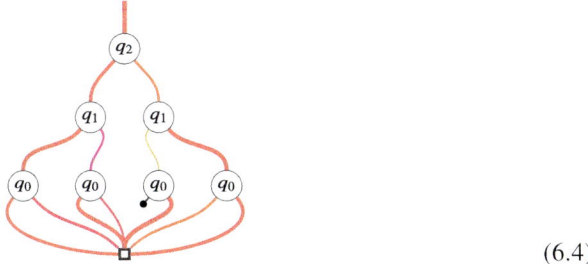

$$(6.4)$$

Since matrix-matrix and matrix-vector multiplication is associative, the order in which the individual multiplications are conducted can, in principle, be chosen arbitrarily. We refer to this order of computations as a *simulation path*. Due to matrix-vector multiplication, in general, being much less complex than matrix-matrix multiplication, the most natural simulation path is to *sequentially* compute the matrix-vector product of individual (and compact) gate matrices with the current state vector. However, for a circuit G with $|G|$ gates, there are

$$|G| * (|G| - 1) * \cdots * 1 = |G|!,\qquad(6.5)$$

i.e., exponentially many, unique simulation paths—raising the question whether the most natural path indeed is always the best path.

In order to demonstrate the impact of the simulation path on the simulation complexity, consider the following typical use case for quantum circuit simulation (cf. Part IV): Given two quantum circuits $G = g_0 \ldots g_{|G|-1}$ and $G' = g'_0 \ldots g'_{|G'|-1}$, it shall be checked whether both circuits are equivalent—an essential question when, e.g., verifying the results of quantum circuit compilation flows. Due to quantum circuits being inherently reversible, this can be checked by concatenating one circuit with the inverse of the other, i.e.,

$$\tilde{G} = GG'^{-1} = g_0 \ldots g_{|G|-1} \, g'^{-1}_{|G'|-1} \cdots g'^{-1}_0,\qquad(6.6)$$

and simulating the resulting circuit with various initial states $|\varphi\rangle$. Whenever G and G' are equivalent, $\tilde{G} \equiv I$ holds (with I denoting the identity transformation), and, hence, $\tilde{G}$ maps $|\varphi\rangle$ to itself. However, as the following example shows, choosing the right simulation path for the simulation of $\tilde{G}$ can make the difference between linear and exponential complexity.

Example 6.2 Consider the scenario as in Example 6.1 and, for the sake of the argument, assume that $G' = G$, i.e., it naturally holds $\tilde{G} = GG^{-1} \equiv I$ for any $|\varphi\rangle$. Then, following the discussion in Example 6.1, simulating $\tilde{G}$ in a sequential fashion leads to an intermediate decision diagram that is maximally large—implying an exponential memory complexity and, hence, exponential runtime. If, however, the simulation path is chosen to start "in between" G and G^{-1} and *alternate* between applying gates from G and G^{-1}, any computation (except the last matrix-vector multiplication) has the form

$$U_i^{-1} * U_i \quad \text{or} \quad I * U_i, \tag{6.7}$$

for $i = 1, \ldots, |G|$. Since, in general, the complexity of decision diagrams representing individual gates is linear, the overall runtime and memory complexity is linear as well. Using a color palette where green denotes and red denotes large decision diagrams, this scenario can be sketched as follows:

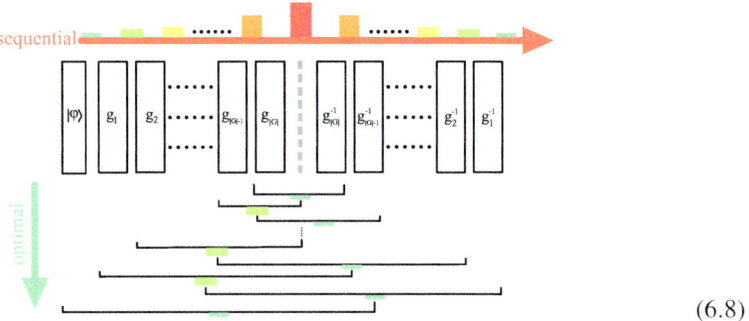

$$\tag{6.8}$$

Obviously, the previous example is specifically constructed to show the extremes—there is a one-to-one correspondence between gates from G and G' and, hence, an easy way to define an "optimal" strategy. In practice, that is, when $G' \neq G$, no such natural correspondence exists, and, as a consequence, it is generally hard to determine an "optimal" strategy for conducting the simulation (cf. Part IV). Although this is hardly surprising, given that equivalence checking of quantum circuits has been shown to be computationally hard [14], it underpins the importance of efficient and automated methods to determine suitable simulation paths.

6.1.2 Related Work

Decision diagrams are not the only data structure for simulating quantum circuits that suffers from the exponential difference in best- and worst-case complexity. As discussed in Sect. 3.2, the performance of tensor network methods [15–17] heavily depends on the order in which the individual tensors are contracted. Both techniques—decision diagrams and tensor networks—efficiently represent the initial state as well as all the individual operations in the form of a dedicated data structure. Then, they choose a certain path to combine these individual descriptions in order to eventually form a representation of the final quantum state—either by multiplying decision diagrams or by contracting tensors. Hence, the problem of determining an optimal simulation path for decision diagrams poses a similar challenge as determining an optimal contraction order for a tensor network.

However, in case of decision diagrams, this question is hardly studied and almost no heuristics exist for determining an efficient simulation path. Initial work related to the problem considered in this work has been conducted in [12, 18]. In [12], it is shown that initially constructing the functionality of certain building blocks in prominent quantum algorithms such as Grover's [19] or Shor's [20] algorithm (by conducting, potentially expensive, matrix-matrix multiplications) can lead to significant runtime improvements compared to the sequential matrix-vector multiplication approach described above. In [18], the authors describe schemes to construct the functionality of such building blocks in a more efficient manner, which can be interpreted as very specific simulation paths. However, still only a significantly limited subset of the immense space of possibilities for very specific problems has been explored.

Besides the sizable difference of research conducted on tensor networks compared to decision diagrams in general, one of the main reasons for this disparity can be identified from the fundamental properties of both techniques: The performance when contracting tensors only depends on their size and shape—not on the actual content (the data) in the tensors. On the one hand, this implies that an a-priori estimate of a particular contraction plan's performance can be efficiently inferred from the sizes and shapes of the tensors involved in all contractions. On the other hand, this also means that without a proper contraction plan, there is nothing to gain by employing tensor networks. In contrast, decision diagrams explicitly try to exploit redundancies in the underlying representations—rather relying on "external" characteristics. This allows them to efficiently represent and simulate even large quantum systems in many cases. Consequently, while tensor networks are in dire need of efficient contraction plans to achieve peak performance, simulating circuits sequentially using decision diagrams has been "good enough" in many cases.

The first steps towards combining both techniques have been taken in [21]. There, the authors use a variation of decision diagrams to represent individual tensors in a more efficient fashion, i.e., they show what can be learned from decision diagrams in order to improve tensor networks. In contrast, this chapter investigates whether knowledge from the tensor network domain can lead to improvements in simulation based on decision diagrams.

6.2   A Simulation Path Framework

In an effort to foster the understanding and development of simulation path heuristics for quantum circuit simulation based on decision diagrams, this section presents an open-source simulation path framework. This framework allows one to execute arbitrary simulation paths using the powerful TaskFlow [22] library. Instead of reinventing the wheel and trying to compensate for years of research on tensor contraction methods, the framework includes a push-button flow to employ existing techniques from the tensor network domain while simultaneously providing the means to easily realize dedicated simulation path strategies for decision diagrams. In the following, we describe how the framework itself handles simulation paths and, afterwards, describe the flow for translating the problem from the domain of decision diagrams to the tensor network domain and back again.

6.2.1   Handling Simulation Paths

The simulation of a quantum circuit $G = g_1 \ldots g_{|G|}$ with the initial state $|\varphi\rangle$ entails the computation of the expression

$$U_{|G|} * \cdots * U_1 * |\varphi\rangle . \tag{6.9}$$

Initially, this requires the construction of decision diagrams for the initial state and the individual gates. Then, each multiplication in the above expression can be regarded as a *task* that takes two decision diagrams and returns the result of their multiplication. Thus, a path for the simulation of G corresponds to a sequence of (multiplication) tasks that eventually result in the final state vector. It is natural to represent such a sequence as a *task dependency graph*. An example illustrates the idea.

Example 6.3 Consider again the three-qubit QFT circuit from Example 6.1. Then, the following sequence of tasks describes one particular simulation path of G:

$$[(0, 1), (2, 3), (4, 5), (6, 7), (8, 9), (10, 11), (12, 13)] \tag{6.10}$$

To this end, index 0 denotes the initial state, index 1 to $|G|$ the individual operations, and the result of a task is indexed by the next largest integer not already in use. This corresponds to the following graph:

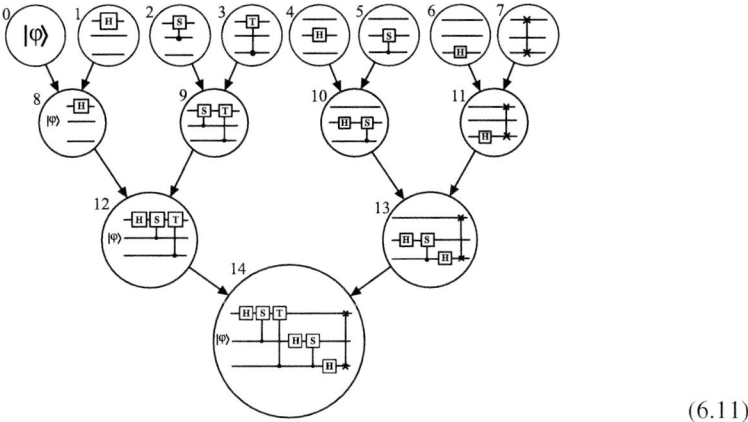

$$(6.11)$$

This task-based formulation of quantum circuit simulation allows one to employ powerful tools for asynchronous task parallelism [22–24] to perform the simulation. The resulting framework takes a circuit and a sequence of tasks as input and uses the Taskflow [22] library to build the corresponding task dependency graph and execute it asynchronously. A similar effort has been conducted for the domain of tensor networks in [25]. The question remains how to determine suitable simulation paths from the $|G|!$ options.

6.2.2   Utilizing Research on Tensor Network Contraction

As reviewed in Sect. 6.1.2, a number of methods have been developed to determine efficient contraction plans for tensor networks, and the first steps have been taken to combine these two techniques. Due to the direct connection between the two domains, it would not make sense to try and reinvent the wheel when it comes to simulation using decision diagrams. Instead, both domains can be connected to make use of research conducted towards tensor network contraction. The following (rather technical) paragraphs give a detailed description of this process.

Starting from an initial quantum circuit (provided in the form of an *OpenQASM* file [26] or Qiskit *QuantumCircuit* object [27]), the first step is to create a corresponding tensor network representation. To this end, each individual gate is transformed into a corresponding tensor representing the underlying matrix. We do not employ tensor slicing techniques (as demonstrated in [3, 28, 29]), which allow splitting the tensors of multi-qubit gates into multiple smaller tensors, since these techniques are not yet widely adopted in the decision diagram domain (although first efforts towards this direction have been conducted in [30], which forms the basis of Chap. 5).

Next, the initial state $|\varphi\rangle$ needs to be translated into the tensor network domain. In general, an n-qubit state is described by a rank-n tensor of size 2^n, that is, the complete state vector. In case of product states, that is, states that can be written as a product of single-qubit states such as the all-zero state $|0\ldots0\rangle = |0\rangle \otimes \cdots \otimes |0\rangle$, this can be represented as n rank-1 tensors of size 2. A similarly compact representation is achieved for decision diagrams of product states, which always consist of n nodes (as opposed to the general worst case of $2^n - 1$ nodes). However, while tensor networks allow for arbitrary contractions between two tensors as long as they share a common index, decision diagrams for quantum computing, as considered in this book, do not support arbitrary kinds of (tensor) contractions (yet). Instead, they only support (proper) matrix-vector and matrix-matrix multiplication, i.e., it is, for example, not possible to contract the (vector) decision diagram representing a single-qubit state and the (matrix) decision diagram representing a two-qubit operation. As a consequence, the initial state in the translated tensor network needs to be represented as a full rank-n tensor (see Sect. 6.3 for further discussion).

Example 6.4 Consider once more the three-qubit QFT circuit from Example 6.1. Then, the corresponding tensor network representation for the simulation of the circuit is given by

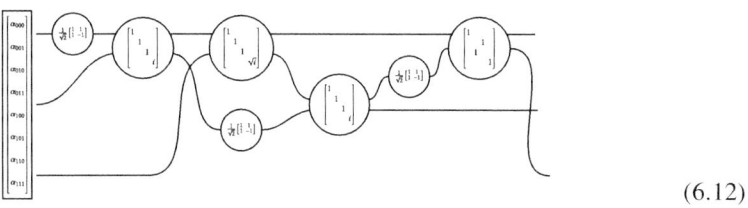

$$(6.12)$$

After the translation, the resulting tensor network can be fed into any available tensor network contraction tool to determine a suitable contraction plan. The hyper-optimized tensor network contraction tool CoTenGra [2] is used as a state-of-the-art representative. It allows one to determine contraction plans for large tensor networks using various graph-based methods and is publicly available at github.com/jcmgray/cotengra. Furthermore, it provides means to visualize contraction plans and their complexity in meaningful ways.

Example 6.5 Feeding the tensor network shown in Example 6.4 into CoTenGra results in the following simulation path:

$$[(0, 1), (2, 8), (3, 9), (4, 10), (5, 11), (6, 12), (7, 13)] \tag{6.13}$$

The resulting contraction tree can be visualized as

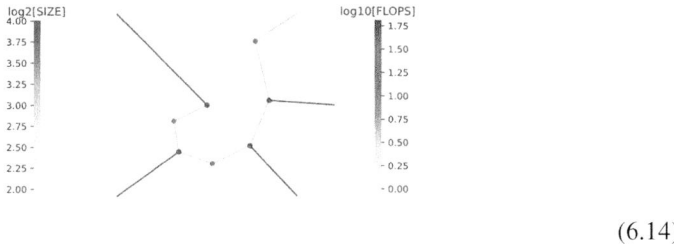

$$(6.14)$$

To this end, the color of each node represents the number of floating point operations required for a particular contraction, while the color of each edge represents the size of the respective tensors—the darker the color, the more complex the contraction or the larger the tensor.

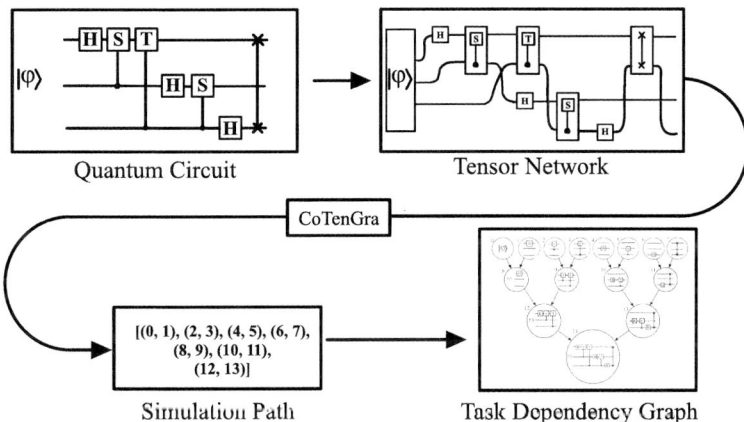

Fig. 6.1 Automated simulation path flow inspired by the tensor network domain

In general, this results in a flow as shown in Fig. 6.1, where the initial quantum circuit is first translated to a tensor network and then fed into CoTenGra. Subsequently, a task dependency graph is constructed from the obtained contraction plan and used for the decision diagram simulation.

6.3   Summary of Results

The proposed simulation path framework has been integrated into the publicly available decision diagram-based simulator MQT DDSIM introduced in Chap. 5 using CoTenGra [2] to determine suitable contraction plans. Inspired by the ideas in [31] (which forms the basis of Chap. 15), a dedicated simulation path heuristic has been developed for the scenario considered in Sect. 6.1. For details on the experimental setup and the numerical results obtained, the reader is referred to [6, 7]. In summary, the following results have been obtained:

- A first series of evaluations using the design flow proposed in Sect. 6.2.2 to make use of the plethora of available tensor network strategies showed that reusing or translating methods developed in the tensor network domain via the proposed flow can already speed up the simulation of quantum circuits using decision diagrams by a large margin compared to the state-of-the-art, i.e., sequential, approach [8].
- Interestingly, in some cases, the substantially improved runtime is overshadowed by the time spent searching for a suitable path. In other cases, nothing is to gain by spending up to 600 s on the search for a suitable path.
- A second series of evaluations on the performance of the dedicated heuristic simulation path heuristic clearly demonstrated the potential of dedicated simulation path schemes for decision diagrams. Although the proposed method is only a heuristic, it often yields several orders of magnitude faster runtimes compared to the state of the art. For some benchmarks, choosing the right simulation path makes the difference between waiting almost two hours for the result and having it available in the blink of an eye.

Although finding a suitable simulation path inspired by the tensor network domain can consume a significant amount of time (in some cases overshadowing the subsequent runtime improvements), the runtime of the simulation paths determined by CoTenGra demonstrates that there *is* something to learn from the domain of tensor networks. However, the current state of the art in decision diagrams for quantum computing, as considered in this work, imposes several limitations on what can actually be learned from the domain of tensor networks. A more detailed discussion on the (dis)similarities of tensor networks and decision diagrams is given in Chap. 7.

Implementation, Usage, Documentation, and Results

 The proposed methodology is available as part of the open-source MQT DDSIM tool at https://github . com/munich-quantum-toolkit/ddsim, which can be installed using pip install mqt . ddsim.

 The resulting tool can be set up as described and illustrated in section 5.3 on page 41. Then, to use the simulation path framework, only the selection of the provider has to be changed, that is,

```
backend = provider.get_backend
                ("path_sim_qasm_simulator")
```

The framework can be configured using multiple options that can be passed to the run function.

 Documentation on all available configuration options is available at https://mqt . readthedocs . io/projects/ddsim

 Details on the experimental setup, evaluations, and results can be found in [6], [7].

References

1. L. Chi-Chung, P. Sadayappan, R. Wenger, On optimizing a class of multi-dimensional loops with reduction for parallel execution. Parallel Process. Lett. (1997)
2. J. Gray, S. Kourtis, Hyper-optimized tensor network contraction. Quantum **5**, 410 (2021)
3. C. Huang, F. Zhang, M. Newman, et al., Classical simulation of quantum supremacy circuits (2020). arXiv: 2005.06787, Preprint
4. S. Boixo, S.V. Isakov, V.N. Smelyanskiy, H. Neven, Simulation of low-depth quantum circuits as complex undirected graphical models (2018). arXiv: 1712.05384, Preprint
5. D. Lykov, R. Schutski, A. Galda, V. Vinokur, Y. Alexeev, Tensor network quantum simulator with step-dependent parallelization (2020). arXiv: 2012.02430, Preprint
6. L. Burgholzer, A. Ploier, R. Wille, Exploiting arbitrary paths for the simulation of quantum circuits with decision diagrams, in *Design, Automation and Test in Europe* (2022)
7. L. Burgholzer, A. Ploier, R. Wille, Simulation paths for quantum circuit simulation with decision diagrams: what to learn from tensor networks, and what not. IEEE Trans. CAD Integr. Circuits Syst. (2022). https://doi.org/10.1109/TCAD.2022.3197969. arXiv: 2203.00703
8. A. Zulehner, R. Wille, Advanced simulation of quantum computations. IEEE Trans. CAD Integr. Circuits Syst. (2019). https://doi.org/10.1109/TCAD.2018.2834427
9. V. Samoladas, Improved BDD algorithms for the simulation of quantum circuits, in *Algorithms - ESA*, ed. by D. Halperin, K. Mehlhorn (2008)
10. G.F. Viamontes, I.L. Markov, J.P. Hayes, High-performance QuIDD-based simulation of quantum circuits, in *Design, Automation and Test in Europe* (2004)
11. S. Hillmich, R. Kueng, I.L. Markov, R. Wille, As accurate as needed, as efficient as possible: approximations in DD-based quantum circuit simulation, in *Design, Automation and Test in Europe* (2020)
12. A. Zulehner, R. Wille, Matrix-vector versus matrix-matrix multiplication: potential in DD-based simulation of quantum computations, in *Design, Automation and Test in Europe* (2019). https://doi.org/10.23919/DATE.2019.8714836
13. T. Grurl, J. Fuß, S. Hillmich, L. Burgholzer, R. Wille, Arrays versus decision diagrams: a case study on quantum circuit simulators, in *Int'l Symposium on Multi-Valued Logic* (2020)
14. D. Janzing, P. Wocjan, T. Beth, "Non-identity check" is QMA-complete. Int. J. Quantum Inform. **03**(03), 463–473 (2005)
15. R. Jozsa, On the simulation of quantum circuits (2006). arXiv: quant-ph/0603163, Preprint
16. I.L. Markov, Y. Shi, Simulating quantum computation by contracting tensor networks. SIAM J. Comput. **38**(3), 963–981 (2008), ISSN: 0097-5397. https://doi.org/10.1137/050644756
17. J.D. Biamonte, V. Bergholm, Tensor networks in a nutshell (2017). arXiv: 1708.00006, Preprint
18. L. Burgholzer, R. Raymond, I. Sengupta, R. Wille, Efficient construction of functional representations for quantum algorithms, in *Int'l Conference of Reversible Computation* (2021)
19. L.K. Grover, A fast quantum mechanical algorithm for database search. Proc. ACM 212–219 (1996)
20. P.W. Shor, Polynomial-time algorithms for prime factorization and discrete logarithms on a quantum computer. SIAM J. Comput. (1997). https://doi.org/10.1137/S0097539795293172
21. X. Hong, X. Zhou, S. Li, Y. Feng, M. Ying, A tensor network based decision diagram for representation of quantum circuits (2020). arXiv: 2009.02618, Preprint
22. T.-W. Huang, D.-L. Lin, C.-X. Lin, Y. Lin, Taskflow: a lightweight parallel and heterogeneous task graph computing system. IEEE Trans. Parallel Distrib. Syst. (2021)
23. H. Carter Edwards, C. R. Trott, and D. Sunderland, Kokkos: Enabling manycore performance portability through polymorphic memory access patterns. J. Parallel Distrib. Comput. **74**(12), 3202–3216

24. H. Kaiser, P. Diehl, A. Lemoine, et al., HPX - The C++ standard library for parallelism and concurrency. JOSS **5**(53), 2352 (2020)
25. T. Vincent, L.J. O'Riordan, M. Andrenkov, et al., Jet: fast quantum circuit simulations with parallel task-based tensor-network contraction (2021). arXiv: 2107.09793, Preprint
26. A.W. Cross, A. Javadi-Abhari, T. Alexander, et al., OpenQASM 3: a broader and deeper quantum assembly language (2021). arXiv: 2104.14722, Preprint
27. A. Javadi-Abhari, M. Treinish, K. Krsulich, et al., Quantum computing with Qiskit (2024). https://doi.org/10.48550/arXiv.2405.08810. arXiv: 2405.08810 [quant-ph]
28. Z.-Y. Chen, Q. Zhou, C. Xue, X. Yang, G.-C. Guo, G.-P. Guo, 64-qubit quantum circuit simulation. Sci. Bull. **63**(15), 964–971 (2018)
29. E. Pednault, J.A. Gunnels, G. Nannicini, et al., Pareto-efficient quantum circuit simulation using tensor contraction deferral (2020). arXiv: 1710.05867, Preprint
30. L. Burgholzer, H. Bauer, R. Wille, Hybrid Schrödinger-Feynman simulation of quantum circuits with decision diagrams, in *Int'l Conference on Quantum Computing and Engineering* (2021). https://doi.org/10.1109/QCE52317.2021.00037
31. L. Burgholzer, R. Raymond, R. Wille, Verifying results of the IBM Qiskit quantum circuit compilation flow, in *Int'l Conference on Quantum Computing and Engineering* (2020). https://doi.org/10.1109/QCE49297.2020.00051

Tensor Networks or Decision Diagrams? Guidelines for Classical Quantum Circuit Simulation

As shown in the previous chapters, decision diagrams (cf. Sect. 3.1) are a promising data structure for efficiently simulating quantum circuits on classical computers. On the other hand, tensor networks (cf. Sect. 3.2) have been heavily used for this task in the literature as well [1–5]. Both data-structures have independently been developed with differing perspectives, terminologies, and backgrounds in mind. This raises the question which is the better data structure for a given use case.

In this chapter (based on [6]), a systematic analysis of the (dis)similarities between tensor networks and decision diagrams for classical quantum circuit simulation is presented. To this end, both data-structures are compared with regard to the following aspects:

- **Abstraction Level**: Several use cases for classically simulating quantum circuits on different abstraction layers exist, e.g., developing/testing of high-level applications or low-level verification of real quantum computers. The question is, where tensor networks and decision diagrams work best.
- **Desired Simulation Output**: The desired output of a classical simulation can be anything from a scalar quantity to the complete state vector. It is key to understand what the limitations of the respective techniques are.
- **Computation Order**: The performance of either technique heavily depends on the order in which the individual operations are conducted. While there is an immediate duality between the task for tensor networks and decision diagrams, the question is, whether the respective techniques are interchangeable between domains.
- **Distributing the Workload**: In order to perform classical simulations of large quantum circuits, the respective methods have to be scalable to supercomputers. This raises the question, how well both techniques allow to make use of the available resources.

© The Author(s), under exclusive license to Springer Nature Switzerland AG 2026
L. Burgholzer and R. Wille, *Design Automation Tools and Software for Quantum Computing*, https://doi.org/10.1007/978-3-032-06770-8_7

Overall, these considerations lead to guidelines for choosing an appropriate data structure to classically simulate quantum circuits depending on the respective use cases.

The rest of this chapter is structured as follows: Sects. 7.1–7.4 compare both techniques according to the aspects introduced above. Based on that, Sect. 7.5 discusses the resulting guidelines.

7.1   Abstraction Level

Several use cases for classically simulating quantum circuits on different abstraction layers exist—encompassing the development or testing of (high-level) applications and the (low-level) verification of quantum computers. Each of these abstraction layers comes with its own characteristics, e.g., various granularities of gate sets or considerations of specific hardware limitations. Motivated by the limited connectivity of many currently available quantum processors, near-term quantum algorithms are generally designed on lower abstraction levels, i.e., circuits typically only contain single- and two-qubit gates and feature rather local interactions. Tensor networks are perfectly suitable in this case due to them capturing the *topological* structure of a quantum circuit. Any such operation only acting on a couple of qubits can be represented by a compact tensor. In contrast, the current state of the art in decision diagram-based simulation requires every decision diagram representing an operation to be extended to the full system size by forming appropriate tensor products with decision diagrams representing the identity. While this yields a dependency on the number of qubits, the resulting decision diagrams are still compact in general—commonly requiring only a linear number of nodes.

Example 7.1 Consider an n-qubit system and assume that a controlled-NOT operation shall be applied with q_{n-1} and q_0 acting as control and target, respectively. The corresponding tensor just consists of the well-known 4×4 matrix representation of the controlled-NOT operation, i.e.,

$$q_{n-1}^{in} - \begin{bmatrix} 1 & 0 & 0 & 0 \\ 0 & 1 & 0 & 0 \\ 0 & 0 & 0 & 1 \\ 0 & 0 & 1 & 0 \end{bmatrix} - q_{n-1}^{out} \\ q_0^{in} \quad\quad\quad\quad\quad\quad\quad\quad\quad q_0^{out}$$

$$(7.1)$$

On the other hand, the extension to the full system size has the form:

$$P_0 \otimes I^{\otimes(n-2)} \otimes I + P_1 \otimes I^{\otimes(n-2)} \otimes X,$$

with $P_i = |i\rangle \langle i|$ denoting the projections on 0 and 1, respectively. The corresponding decision-diagram representation consists of $2n - 1$ nodes:

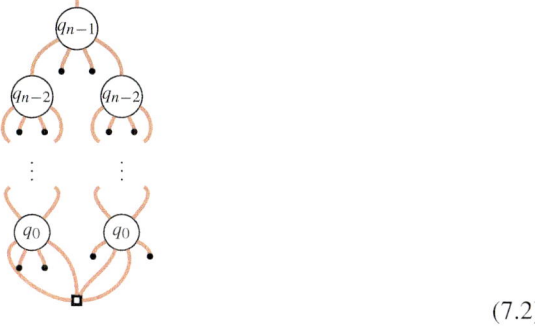

$$(7.2)$$

Similar to the way classical programming languages allow to abstract from assembly or machine code, higher levels of abstractions are used for developing quantum algorithms. Descriptions of such algorithms comprise building blocks such as (Boolean) oracles, basis transformations, or problem encodings. While the functionality of such building blocks typically possesses some kind of structure, this structure is not necessarily topological or local. Since any operation acting on k qubits is inherently represented by a $2^k \times 2^k$ matrix, methods based on tensor networks cannot represent these constructs efficiently for large k. In contrast, decision diagrams aim to exploit *structural redundancies* in the underlying representations which frequently allows them to efficiently represent high-level, non-local circuit elements.

Example 7.2 Consider the simulation of the famous Grover algorithm, which provides a quadratic speed-up over classical techniques for unstructured search problems [7]. An essential part of Grover's algorithm is an *oracle* that encodes solutions to the considered problem. Such an oracle "marks" desired states and is described by a unitary matrix U_f that acts in the following way:

$$U_f \, |x\rangle \otimes |0\rangle = \begin{cases} |x\rangle \otimes |1\rangle & \text{if } x \text{ is the state searched for,} \\ |x\rangle \otimes |0\rangle & \text{otherwise.} \end{cases} \qquad (7.3)$$

One of the simplest cases is the search for the all-one state $|1 \ldots 1\rangle$. An oracle for this task is given by a multi-controlled NOT (or *Toffoli*) gate which flips the value of a target qubit only if all of its control qubits are in state $|1\rangle$. The corresponding unitary matrix U_f is given by $diag(\mathbb{I}, \ldots, \mathbb{I}, X)$. As a consequence, the memory required to store this matrix as a tensor grows exponentially with the system size. In contrast, a linear number of nodes suffices to represent it as a decision diagram:

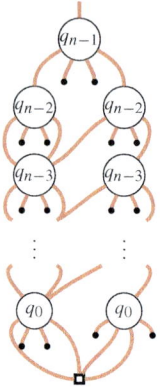

$$(7.4)$$

Conclusion: Tensor networks fit well for low-level applications of classical circuit simulation due to them capitalizing on inherent *topological structure*. In contrast, decision diagrams exploit *structural redundancies* in the underlying representation, which is independent of the abstraction level. As a result, decision diagrams show clear benefits for high-level use cases while, at the same time, covering low-level applications.

7.2   Desired Simulation Output

The desired output of a classical quantum circuit simulation can be anything from a scalar quantity (e.g., an individual amplitude) to the complete state vector. In the following, the implications of the desired quantity on the performance of classical simulation methods based on tensor networks and decision diagrams are studied.

First, assume that a representation of the complete output state vector of a quantum circuit shall be computed. Using tensor networks, this amounts to contracting the whole tensor network into a single rank-n tensor of size 2^n. Inevitably, this incurs the same exponential memory requirement as for the straight-forward matrix-vector calculation. As a consequence, it is infeasible in general to compute the complete output state vector (or, more generally, exponentially many amplitudes) using tensor networks. Many specialized types of tensor networks, such as *Matrix Product States* (MPS), *Tree Tensor Networks* (TTN), *Multi-scale Entanglement Renormalization Ansatz* (MERA), or *Projected Entangled Pair States* (PEPS), have been proposed that try to impose some structure in the whole state representation by decomposing it into smaller tensors (see [8, 9] and the references therein). However, these techniques typically aim for a slightly different goal than considered in this work—namely, efficiently *approximating* the state of quantum mechanical systems.

Using decision diagrams, individual quantum operations can commonly be represented in linear size (as discussed in the previous section). Computing a representation of the full output state then entails the subsequent multiplication of the respective decision diagrams.

It has been shown, that, for many practically relevant cases, the size of the (intermediate) diagrams during the simulation remains polynomial with respect to the number of qubits [10]. Consequently, the inherent exponential complexity is alleviated in those cases. Once the final decision diagram is obtained, an arbitrary number of amplitudes can be extracted with little overhead by recursively traversing the decision diagram and accumulating edge weights along the way.

On the other end of the spectrum, it might be desirable to determine a single scalar quantity, e.g., an individual amplitude or the expected value of some observable. For methods based on tensor networks, this is accomplished by fixing the output indices of the circuit's tensor network, as exemplary illustrated by

$$
\begin{array}{c}
|0\rangle \!-\!\boxed{}\!-\!\langle 0| \\[-4pt]
\quad\;\; U \qquad \xrightarrow{contract} \;\; \alpha_{00} \\[-4pt]
|0\rangle \!-\!\boxed{}\!-\!\langle 0|
\end{array}
\tag{7.5}
$$

In contrast to the calculation of the complete output state, contracting the resulting tensor network results in a single rank-0 tensor, i.e., a scalar. Whenever a contraction plan can be employed that keeps the size and bond dimension of intermediate tensors in check, individual amplitudes can be determined very efficiently. At the moment, the only established way to determine individual amplitudes using decision diagrams is to compute the decision diagram for the full state vector and, then, extract the desired amplitude by traversing the decision diagram from top to bottom. While there still are many instances where this procedure works just fine, this still imposes a significant restriction.

Conclusion: While a general conclusion cannot be drawn without prior assumptions, the following tendency stands out: Tensor networks should be preferred for computing scalar quantities as the result of a quantum circuit simulation—given that a suitable contraction plan can be determined. Decision diagrams are more suited towards full state vector simulation—provided that the involved decision diagrams remain compact throughout the simulation.

7.3   Determining the Order of Computation

Given a particular quantum circuit simulation, there exists an immediate duality between the contraction plan for tensor networks, i.e., the order in which the tensors are contracted, and the simulation path for decision diagrams, i.e., the sequence in which the individual multiplications are performed. Both techniques have in common that their performance heavily depends on the order in which the individual operations are conducted. This problem has been intensively studied for tensor networks [1–4, 11]. While not as extensive as for tensor networks, research towards this direction has also been conducted for decision diagrams [12–16]—in particular, for the use case of equivalence checking (cf. Part IV). Thus, in the following, we consider equivalence checking as the perfect use case for answering the question, whether the strategies for determining the order of computation are interchangeable between domains. This further adds to the finding in Chap. 6.

To this end, given two quantum circuits G and G', it shall be decided, whether simulating both circuits yields equivalent outputs for a fixed input $|\varphi\rangle$. One possible solution is to simulate both circuits with input $|\varphi\rangle$ and compare the resulting states. However, in general, this is neither feasible for tensor networks (due to the inherent exponential memory requirement independent of the contraction plan), nor for decision diagrams (due to their worst case exponential complexity independent of the simulation path). Due to the reversible nature of quantum computing, there is a more promising approach for checking the equivalence of both circuits (cf. Chap. 14). To this end, one of the circuits is inverted and concatenated with the other circuit—forming a new circuit $\tilde{G} = G\, G'^{-1}$. Whenever both original circuits are indeed equivalent for input $|\varphi\rangle$, simulating the circuit $\tilde{G}$ with input $|\varphi\rangle$ does not affect the state at all, i.e., it holds that $|\langle\varphi|\tilde{G}|\varphi\rangle|^2 = 1$. Now, the result of the computation is not a (potentially) exponential representation, but a single complex number. As a consequence, the order of operations to arrive at this compact result becomes crucial.

Due to the topological structure of tensor networks, their performance is independent of the actual content of the individual tensors, but only depending on their size and shape. As a result, precise estimates for the required amount of memory and floating point operations for contracting a tensor network according to a particular contraction plan can be derived a-priori, i.e., without conducting the actual contraction. This makes it possible to explore the vast search space of contraction plans with the goal of keeping the dimension of shared indices and the sizes of the intermediate tensors during the contraction as small as possible. However, determining an optimal contraction plan is an extremely delicate process, and has even been shown to be NP-hard [11]. Accordingly, a plethora of heuristic methods have been proposed to find efficient contraction plans for tensor networks [1–4]. In general, these techniques start from the "outside", i.e., the small tensors of the initial state and the output amplitudes, and "work their way inwards" with the goal of keeping the size of intermediate tensors in check. While estimating the complexity of the multiplication of two decision diagrams is straight-forward (it scales with the product of the decision diagrams' nodes), estimating the size of the resulting decision diagram is extremely difficult in the general case without performing the actual multiplication (due to decision diagrams heavily relying on the amount of redundancy being present in the resulting representation). As a result, no clear runtime estimation can be derived for classical simulation based on decision diagrams. However, it has been shown in [15, 16] (which form the basis of Chap. 6), that translating tensor network contraction plans to simulation paths for decision diagrams can allow for speedups of up to several orders of magnitude compared to the established simulation approach.

The efficiency of a simulation path for decision diagrams heavily depends on the amount of redundancy present in the underlying representations of intermediate results. While not as extensive as for tensor networks, it has been shown that choosing the right simulation path can make the difference between linear and exponential runtime as well as space [12–15]. The resulting strategies mainly achieve their efficiency from incorporating certain characteristics about the considered simulation task. For the problem at hand, the quintessence is to try to keep the decision diagrams occurring during the computation as close as possible to the identity by starting "in between" G and G'^{-1} and *alternating* between applications of gates from G and (inverted) gates from G' according to some scheme. Whenever a scheme, i.e., a simulation path, can be employed that allows to keep the intermediate decision diagrams close to the identity, the equivalence of both circuits can be concluded efficiently. While the same strategy could be employed to contract the corresponding tensor network, this will most certainly not result in an efficient contraction plan. This is due to the fact that, sooner or later, all qubits will be involved in the computation—leading to a maximally large tensor of size $2^n \times 2^n$. Therefore, the opposite direction, i.e., translating a simulation path to a corresponding contraction plan, can hardly be expected to be effective.

Conclusion: Contraction plan and simulation path seem very similar from an outside perspective. But, when taking a closer look, it is clear that they are trying to achieve different goals. An efficient contraction plan tries to keep the size and shape of intermediate tensors in check. In contrast, the goal of simulation paths is to have intermediate decision diagrams which are as compact as possible. While translating an efficient contraction plan for tensor networks into a simulation path has been demonstrated to yield speedups of up to several orders of magnitude, the inverse, i.e., translating an efficient simulation path for decision diagrams into a contraction plan, can hardly be expected to be effective.

7.4   Distributing the Workload

In order to perform classical simulations of large quantum circuits, the respective methods have to be designed to run on high performance supercomputing clusters. Hence, it is key to distribute the workload of a particular computation to multiple (heterogeneous) "workers" in order to fully utilize the available resources. To this end, the task of contracting two tensors is inherently parallelizable on CPUs as well as GPUs due to its regular structure. On the other hand, it has been shown in [17] that parallelizing individual decision-diagram operations is not straight-forward (not even on CPUs). Intuitively, the fashion in which decision diagrams try to exploit redundancies mitigates the potential of trivial parallelization as it exists for regular matrices and vectors. Given a specific contraction plan, independent contractions can be efficiently conducted in parallel [18]. Due to the immediate duality between contraction plans and simulation paths witnessed in the previous section, this also holds for decision diagrams. However, some additional effort is required to concurrently maintain multiple decision-diagram packages. Furthermore, methods based on tensor networks employ *slicing* or *cutting* techniques, which essentially subdivide a tensor network by fixing the values of certain indices [1, 19, 20]. There, the main idea is to split multi-qubit tensors into multiple smaller tensors of higher order, as exemplary illustrated in the following:

$$\begin{array}{c} |0\rangle \\ |0\rangle \end{array} \boxed{U} \;\;=\;\; \begin{array}{cc} |0\rangle & \boxed{U_1} \\ |0\rangle & \boxed{U_0} \end{array}. \tag{7.6}$$

While first efforts towards employing such techniques for decision diagrams have been conducted in [21] (which forms the basis of Chap. 5), they are not yet mature and flexible enough to be employed in the same way as for tensor networks. This is due to one of the main (technical) restrictions of decision diagrams *at the moment*: They only allow for proper matrix-vector and matrix-matrix multiplication while tensor networks allow for arbitrary contractions between two tensors (as long as they share a common index). So, while the following contraction between a single-qubit state and a two-qubit operation makes perfect sense for tensor networks

$$\begin{array}{c} |0\rangle \\ |0\rangle \end{array} \boxed{U} \;\;=\;\; |0\rangle \boxed{U'}, \tag{7.7}$$

decision diagrams currently only permit the following type of contraction

$$\begin{array}{c} |0\rangle \\ \otimes \\ |0\rangle \end{array} \boxed{U} \;\;=\;\; \boxed{U \,|00\rangle}. \tag{7.8}$$

Conclusion: Significant efforts have been conducted to scale methods based on tensor networks to the realm of exascale computing [18, 22, 23]. In contrast, corresponding solutions for decision diagrams are still in their infancy and the current state of the art is rather tailored for desktop systems—its limits yet to be explored. As a consequence, classical computations near the quantum supremacy threshold will probably be tackled using tensor networks for the foreseeable future, while decision diagrams are a suitable means to classically simulate quantum circuits on desktop computers, e.g., during application development.

7.5   Summary of Results

In this chapter, two complementary approaches to tackle the inherent complexity of classically simulating quantum circuits have been systematically analyzed—tensor networks and decision diagrams. To this end, both techniques were compared with regard to their most

	Tensor Networks	Decision Diagrams
Abstraction level	Efficient for low-level applications through exploiting topological structure	Independently applicable through exploiting structural redundancies with clear benefits for high-level applications
Desired simulation output	Should be preferred for computing scalar quantities given a suitable contraction plan can be determined	Should be preferred for computing the full state vector given that the involved decision diagrams remain compact
Computation order	A plethora of methods is available aiming to keep the size and shape of the tensors in check	First methods have been developed aiming to keep intermediate decision diagrams as compact as possible
Distributing the workload	Vast options are available for simulations near the supremacy threshold on HPC systems	Potential has been demonstrated, but solutions beyond desktop computers are yet to be explored

Fig. 7.1 Guidelines for choosing between tensor networks and decision diagrams

applicable *abstraction level*, the *desired simulation output*, the impact of the *computation order*, and the ease of *distributing the workload*. This results in the guidelines for choosing either decision diagrams or tensor networks for a particular use case shown in Fig. 7.1.

References

1. J. Gray, S. Kourtis, Hyper-optimized tensor network contraction. Quantum **5**, 410 (2021)
2. C. Huang, F. Zhang, M. Newman, et al., Classical simulation of quantum supremacy circuits (2020). arXiv: 2005.06787, Preprint
3. S. Boixo, S.V. Isakov, V.N. Smelyanskiy, H. Neven, Simulation of low-depth quantum circuits as complex undirected graphical models (2018). arXiv: 1712.05384, Preprint
4. D. Lykov, R. Schutski, A. Galda, V. Vinokur, Y. Alexeev, Tensor network quantum simulator with step-dependent parallelization (2020). arXiv: 2012.02430, Preprint
5. J.D. Biamonte, V. Bergholm, Tensor networks in a nutshell (2017). arXiv: 1708.00006, Preprint
6. L. Burgholzer, A. Ploier, R. Wille, Tensor networks or decision diagrams? guidelines for classical quantum circuit simulation (2023). arXiv:2302.06616 [quant-ph], Preprint
7. L.K. Grover, A fast quantum mechanical algorithm for database search. Proc. ACM 212–219 (1996)
8. G.F. Viamontes, I.L. Markov, J.P. Hayes, Checking equivalence of quantum circuits and states, in *Int'l Conference on CAD* (2007)
9. I. Cirac, D. Perez-Garcia, N. Schuch, F. Verstraete, Matrix product states and projected entangled pair states: concepts, symmetries, and theorems (2021). arXiv: 2011.12127, Preprint
10. A. Zulehner, R. Wille, Advanced simulation of quantum computations. IEEE Trans. CAD Integr. Circuits Syst. (2019). https://doi.org/10.1109/TCAD.2018.2834427
11. L. Chi-Chung, P. Sadayappan, R. Wenger, On optimizing a class of multi-dimensional loops with reduction for parallel execution. Parallel Process. Lett. (1997)
12. A. Zulehner, R. Wille, Matrix-vector versus matrix-matrix multiplication: potential in DD-based simulation of quantum computations, in *Design, Automation and Test in Europe* (2019). https://doi.org/10.23919/DATE.2019.8714836
13. L. Burgholzer, R. Wille, Advanced equivalence checking for quantum circuits. IEEE Trans. CAD Integr. Circuits Syst. (2021). https://doi.org/10.1109/TCAD.2020.3032630
14. L. Burgholzer, R. Raymond, R. Wille, Verifying results of the IBM Qiskit quantum circuit compilation flow, in *Int'l Conference on Quantum Computing and Engineering* (2020). https://doi.org/10.1109/QCE49297.2020.00051

15. L. Burgholzer, A. Ploier, R. Wille, Exploiting arbitrary paths for the simulation of quantum circuits with decision diagrams, in *Design, Automation and Test in Europe* (2022)
16. L. Burgholzer, A. Ploier, R. Wille, Simulation paths for quantum circuit simulation with decision diagrams: what to learn from tensor networks, and what not. IEEE Trans. CAD Integr. Circuits Syst. (2022). https://doi.org/10.1109/TCAD.2022.3197969. arXiv: 2203.00703
17. S. Hillmich, A. Zulehner, R. Wille, Concurrency in DD-based quantum circuit simulation, in *Asia and South Pacific Design Automation Conference* (2020). https://doi.org/10.1109/ASP-DAC47756.2020.9045711
18. T. Vincent, L.J. O'Riordan, M. Andrenkov, et al., Jet: fast quantum circuit simulations with parallel task-based tensor-network contraction (2021). arXiv: 2107.09793, Preprint
19. B. Villalonga, S. Boixo, B. Nelson, et al., A flexible high-performance simulator for verifying and benchmarking quantum circuits implemented on real hardware. NPJ Quantum Inf. (2019). ISSN: 2056-6387. https://doi.org/10.1038/s41534-019-0196-1
20. Z.-Y. Chen, Q. Zhou, C. Xue, X. Yang, G.-C. Guo, G.-P. Guo, 64-qubit quantum circuit simulation. Sci. Bull. **63**(15), 964–971 (2018)
21. L. Burgholzer, H. Bauer, R. Wille, Hybrid Schrödinger-Feynman simulation of quantum circuits with decision diagrams, in *Int'l Conference on Quantum Computing and Engineering* (2021). https://doi.org/10.1109/QCE52317.2021.00037
22. T. Nguyen, D. Lyakh, E. Dumitrescu, D. Clark, J. Larkin, A. McCaskey, Tensor network quantum virtual machine for simulating quantum circuits at exascale (2021). arXiv: 2104.10523, Preprint
23. J. Brennan, M. Allalen, D. Brayford, et al., Tensor network circuit simulation at exascale, in *International Workshop on Quantum Computing Software (QCS)* (IEEE, 2021), pp. 20–26. ISBN: 978-1-72818-674-0. https://doi.org/10.1109/QCS54837.2021.00006

Summary of Part II

This part of the book has explored the challenge of classically simulating quantum circuits, which is an extremely important and, at the same time, incredibly complex task in the development of quantum computing applications. Efficient data structures such as tensor networks and decision diagrams have been proposed to tackle the resulting complexity. Compared to tensor networks, decision diagrams are still a very young data structure. As such, there is much to learn about their use in classical quantum circuit simulation. More precisely, this part of the book presented the following contributions:

- It was demonstrated that a hybrid Schrödinger-Feynman simulation scheme cannot only be employed using arrays or tensor networks, but also using decision diagrams (cf. Chap. 5). Due to the substantially decreased complexity of the individual simulations, decision diagrams are employed in a regime where more redundancy can be exploited. By handing off the accumulation of individual results to computations on arrays, the bottleneck of decision-diagram addition observed in this work can be effectively circumvented. The resulting scheme combines the best of both worlds and allows one to significantly advance the state of the art in decision diagram-based quantum circuit simulation and, for the first time, allows one to fully exploit the available hardware resources.
- Then, the importance of the path that is chosen when simulating quantum circuits using decision diagrams was studied (cf. Chap. 6). The resulting framework allows one to employ arbitrary simulation paths and to connect the domain of tensor networks with the domain of decision diagrams. It was demonstrated that much can be learned from the domain of tensor networks and that the development of application-specific heuristics that are tailored for decision diagrams can lead to even further improvements.

L. Burgholzer and R. Wille, *Design Automation Tools and Software for Quantum Computing*, https://doi.org/10.1007/978-3-032-06770-8_8

- Finally, it was shown that decision diagrams and tensor networks differ in some key aspects and that not everything can be learned from tensor networks (cf. Chap. 7). Instead, both data structures rather serve as complementary alternatives that each possess their own advantages and disadvantages. This resulted in guidelines that aid designers in deciding which data structure to choose for which use cases.

All the efforts listed above have been integrated into the open-source classical quantum circuit simulator DDSIM, which is available at https://github.com/munich-quantum-toolkit/ ddsim. All methods are provided in a push-button fashion that is easily accessible and extendable.

In today's digital world, creating computer programs has become a crucial element of software development. With the advent of high-level programming languages such as C++ or Python, the development process has become simpler and more efficient. These languages enable developers to produce code that is more human-readable and understandable without having to worry about the underlying hardware's low-level features. But before these programs can be executed on a computer, they must be translated into machine code that the computer can process. This procedure is known as *compilation*, and it entails converting high-level code into a binary format that the computer's processor can directly execute. By making it easier for more people to create computer programs, this has enabled the development of complex software applications that can run on many different platforms.

Just as in classical computing, the design of quantum circuits and the development of quantum algorithms are fundamental in the development of quantum computing applications. Quantum circuits are analogous to classical functions or programs in that they are a sequence of quantum gates that perform specific operations on quantum bits or qubits. Similarly to classical processors, quantum processors can only execute a certain set of native instructions, and they might further limit the qubits on which these operations might be applied. Thus, any high-level quantum circuit (describing a quantum algorithm) must be *compiled* into a representation that can be executed on the targeted device. Most importantly, the resulting quantum circuit must only use gates that are native to the device on which it shall be executed. If the device only has limited connectivity between its qubits, it must only apply gates to qubits that are connected on the device. Naturally, the efficiency of this compilation process is critical because it can have a significant impact on the performance of the resulting quantum program. Inefficient compilation can lead to longer execution times, higher error rates, and reduced accuracy in the final result. Therefore, developing efficient compilation methods for quantum programs is essential to overcome the challenges of quantum computing and realize the potential of this technology.

In this part of the book, particularly, the *quantum circuit mapping* task is considered. This is a crucial step in the compilation flow, as it directly affects the feasibility and performance of the quantum circuit on a given device. It involves finding a way to map the qubits of a quantum circuit to the qubits of a quantum device, while respecting the limited connectivity constraints of the device and minimizing the overhead of additional gates. To this end, a method using powerful reasoning engines is proposed (based on [1]) that allows one to determine the minimal amount of additional gates required to *map* a given quantum circuit to a certain quantum computer that limits the interactions between its qubits. Then, based on [2], it is shown how the search space of the optimal mapping problem for quantum circuits can be limited to significantly improve the performance of the mapping scheme.

To familiarize the reader with the fundamental principles, the remainder of this chapter presents an outline of the compilation flow of quantum circuits with a particular focus on mapping.

9.1   The Quantum Circuit Compilation Flow

Today's quantum computers only support a very limited set of quantum operations natively. In many cases, this native gate set is just large enough to be universal for quantum computing. Typically, it consists of a collection of single-qubit gates and at least one particular two-qubit entangling gate.

Example 9.1 The following provides a list of some common native gate sets among existing quantum computing platforms:

- *IBM (until 2021):* $[U_3(\theta, \phi, \lambda), CX]$,
- *IBM:* $[Id, R_Z(\theta), \sqrt{X}, X, CX]$,
- *Rigetti:* $[R_X(\theta), R_Z(\theta), CZ, CP(\lambda), R_{XX+YY}(\theta, \beta)]$,
- *OQC:* $[R_Z(\theta), \sqrt{X}, X, ECR]$,
- *IonQ/AQT:* $[R_X(\theta), R_Y(\theta), R_Z(\theta), R_{XX}(\theta)]$.

The remainder of this part of the book mainly focuses on the first gate set (that is, arbitrary single-qubit gates and CNOTs). However, all findings remain general and can also be applied to other native gate sets.

As a consequence of the limited gate set, any non-native operation in a quantum circuit has to be *decomposed* into a sequence of native operations [3–7]—frequently referred to as *synthesis*. Most importantly, none of the quantum computing platforms listed in Example 9.1 supports gates that act on more than two qubits. Thus, any gate acting on more than

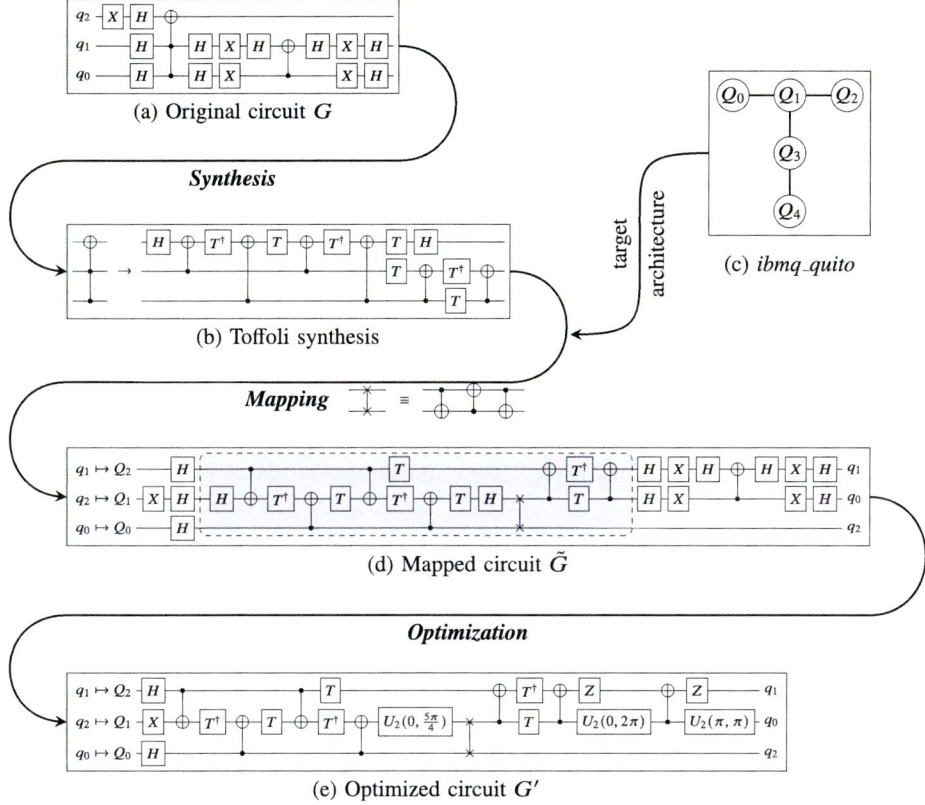

Fig. 9.1 a–e Exemplary illustration of the quantum circuit compilation flow

two qubits must be broken down into (native) single- and two-qubit gates. This process may require the use of additional *ancillary* qubits for realizing the desired operation. Typically, there is a trade-off between the number of gates required for the decomposition of multi-qubit gates and the number of additional qubits being used—the more qubits available, the fewer gates are required.

Example 9.2 Consider the quantum circuit shown in Fig. 9.1a. It represents a small instance of Grover's algorithm. For the algorithm to run on any of the platforms listed in Example 9.1, the three-qubit Toffoli gate has to be decomposed/synthesized into the corresponding single- and two-qubit gates. Figure 9.1b shows an exemplary decomposition of the Toffoli gate in the Clifford+T gate-set.

In addition to having a limited set of gates, many existing quantum computers also limit the pairs of qubits that can interact with each other directly. This is usually described by a *coupling graph* (V, E), where the nodes of the graph (V) represent the qubits, and an edge between two nodes (E) indicates that a two-qubit operation may be applied to those qubits. In the past, these coupling graphs typically were directed, i.e., the order of the qubits in a two-qubit gate mattered (e.g., which qubit acts as the control qubit and which acts as the target qubit in a CNOT gate). The main reason for this was that physical realizations of quantum computers often show a preferred direction for two-qubit interactions, which reduces error rates. Since then, most companies that offer quantum computers have chosen to abstract this effect away and describe their systems as bi-directional. In general, this makes the architectures more connected, which makes it easier to meet the limited connectivity requirements. At the same time, it ignores the underlying physical preference for certain directions, so it might be harder to use the respective device to its full potential.

Example 9.3 Consider the coupling graph shown in Fig. 9.1c. It describes the architecture of the five-qubit, T-shaped *ibmq_quito* device. Any two-qubit gate that will be executed on this device must be applied to one of (Q_0, Q_1), (Q_1, Q_2), (Q_1, Q_3), or (Q_3, Q_4). Since the coupling graph is bidirectional, the order of the qubits involved in the gate does not matter. For example, a CNOT might be applied with Q_0 as a control and Q_1 as a target, but also vice versa.

As a consequence of the limited connectivity, realizing a generic quantum algorithm on such a device requires a *mapping* step that maps the qubits of the circuit to the qubits of the device (commonly referred to as *qubit layout* or *qubit allocation*) and ensures that any gate acts only on qubits that are connected on the device (commonly referred to as *qubit routing*). In the literature, the qubits of the circuit are commonly called *logical qubits*, while the qubits of the device are called *physical qubits*. These notions are merely used to denote different abstraction levels and should not be confused with a similar terminology from quantum error correction, where multiple physical qubits form a single logical qubit that is protected from errors.

In most cases, it is not possible to statically define a mapping of the circuit's qubits to the device's qubits such that all gates of the circuit conform to the connectivity limitations of the device. Consequently, this mapping has to change dynamically throughout the circuit. This can be accomplished by using *SWAP* gates that allow the position of two logical qubits on the architecture to be interchanged.

Example 9.4 Consider again the circuit G from Example 9.2 and assume that the Toffoli gate has been synthesized as shown in Fig. 9.1b. Furthermore, assume that the circuit is to be executed on the *imbq_quito* architecture shown in Fig. 9.1c. Then Fig. 9.1d shows one possible circuit $\tilde{G}$ resulting from this mapping process. The physical qubits Q_0, Q_1, and Q_2 were chosen and initially assigned logical qubits q_0, q_2, and q_1, respectively. Only one *SWAP* operation was applied to Q_0 and Q_1 (indicated by $\times$) in the middle of the circuit to conform to the connectivity constraints of the target.

Although many (heuristic) techniques have been proposed in the past that allow one to determine suitable mappings, e.g., [8–16], determining truly optimal solutions (with as little overhead as possible) revealed to be a challenging problem. In fact, the mapping problem has been shown to be NP-complete [17, 18]. To this end, Chap. 10 proposes a technique that guarantees solutions with the minimum number of additional gates.

While any synthesized and mapped circuit may be executed on a targeted device, it is not guaranteed to produce meaningful results if the circuit, e.g., is too large for the state to stay coherent. This is due to the qubits of a device inherently being affected by noise—leading to rather *short coherence times* and *limited fidelity* of the individual operations. Until a certain threshold concerning the number of available qubits is reached, error correction is not yet an option. On the one hand, this motivates quantum circuit optimizations, such as gate fusion, gate cancellation, or blockwise re-synthesis—which aim to reduce the overall gate count of circuits to be executed in order to reduce the effect of noise and allow the computation to stay coherent [19–27]. On the other hand, mapping techniques that take into account the calibration and error data of the targeted device to achieve noise-adaptive mappings shift into focus [12, 28].

Example 9.5 The simplest optimization scheme is the *single-qubit gate fusion*. Since any single-qubit unitary represents a rotation of the Bloch sphere, any sequence of such gates also represents a rotation. Instead of performing each rotation separately, the complete rotation can be performed at once. Now, consider again the circuit $\tilde{G}$ from Example 9.4 shown in Fig. 9.1d that has been mapped to the *ibmq_quito* architecture. Then, nine single-qubit gates can be eliminated by this optimization—resulting in the *optimized* circuit G' shown in Fig. 9.1e.

References

1. R. Wille, L. Burgholzer, A. Zulehner, Mapping quantum circuits to IBM QX architectures using the minimal number of SWAP and H operations, in *Design Automation Conference* (2019). https://doi.org/10.1145/3316781.3317859

2. L. Burgholzer, S. Schneider, R. Wille, Limiting the search space in optimal quantum circuit mapping, in *Asia and South Pacific Design Automation Conference* (2022). https://doi.org/10.1109/ASP-DAC52403.2022.9712555

3. B. Giles, P. Selinger, Exact synthesis of multiqubit Clifford+T circuits. Phys. Rev. A **87**(3), 032–332 (2013). ISSN: 1050-2947, 1094-1622. https://doi.org/10.1103/PhysRevA.87.032332

4. M. Amy, D. Maslov, M. Mosca, M. Roetteler, A meet-in-the-middle algorithm for fast synthesis of depth-optimal quantum circuits. IEEE Trans. CAD Integr. Circ. Syst. **32**(6), 818–830 (2013). ISSN: 0278-0070, 1937-4151. https://doi.org/10.1109/TCAD.2013.2244643

5. R. Wille, M. Soeken, C. Otterstedt, R. Drechsler, Improving the mapping of reversible circuits to quantum circuits using multiple target lines, in *Asia and South Pacific Design Automation Conference* (2013)

6. D.M. Miller, R. Wille, Z. Sasanian, Elementary quantum gate realizations for multiple-control Toffoli gates, in *International Symposium on Multi-Valued Logic* (2011). ISBN: 978-1-4577-0112-2. https://doi.org/10.1109/ISMVL.2011.54

7. D. Maslov, On the advantages of using relative phase Toffolis with an application to multiple control Toffoli optimization. Phys. Rev. A **93**(2), 022–311 (2016). ISSN: 2469-9926, 2469-9934. https://doi.org/10.1103/PhysRevA.93.022311

8. A. Zulehner, A. Paler, R. Wille, An efficient methodology for mapping quantum circuits to the IBM QX architectures. IEEE Trans. CAD Integr. Circ. Syst. (2019). https://doi.org/10.1109/TCAD.2018.2846658

9. K.N. Smith, M.A. Thornton, A quantum computational compiler and design tool for technology-specific targets, in *International Symposium on Computer Architecture* (2019), pp. 579–588

10. G. Li, Y. Ding, Y. Xie, Tackling the qubit mapping problem for NISQ-era quantum devices, in *International Conference on Architectural Support for Programming Languages and Operating Systems* (2019). https://doi.org/10.1145/3297858.3304023

11. A. Matsuo, W. Hattori, S. Yamashita, Reducing the overhead of mapping quantum circuits to IBM Q system, in *IEEE International Symposium on Circuits and Systems* (2019)

12. P. Murali, J.M. Baker, A. Javadi-Abhari, F.T. Chong, M. Martonosi, Noise-adaptive compiler mappings for noisy intermediate-scale quantum computers, in *International Conference on Architectural Support for Programming Languages and Operating Systems* (ACM, Providence RI USA, 2019), pp. 1015–1029. ISBN: 978-1-4503-6240-5. https://doi.org/10.1145/3297858.3304075

13. M. Amy, V. Gheorghiu, Staq—A full-stack quantum processing toolkit. Quant. Sci. Technol. **5**(3), 034–016 (2020)

14. S. Sivarajah, S. Dilkes, A. Cowtan, W. Simmons, A. Edgington, R. Duncan, T—ket: a retargetable compiler for NISQ devices. Quant. Sci. Technol. **6**(1), 014–003 (2021). ISSN: 2058-9565. https://doi.org/10.1088/2058-9565/ab8e92

15. Y. Hirata, M. Nakanishi, S. Yamashita, Y. Nakashima, An efficient conversion of quantum circuits to a linear nearest neighbor architecture. Quant. Inf. Comput. **11**, 142–166 (2011). https://doi.org/10.26421/QIC11.1-2-10

16. A. Zulehner, R. Wille, Compiling SU(4) quantum circuits to IBM QX architectures, in *Asia and South Pacific Design Automation Conference* (2019), pp. 185–190. https://doi.org/10.1145/3287624.3287704

17. A. Botea, A. Kishimoto, R. Marinescu, On the complexity of quantum circuit compilation, in *International Symposium on Combinatorial Search* (2018)

18. M.Y. Siraichi, V.F. dos Santos, C. Collange, F.M.Q. Pereira, Qubit allocation, in *International Symposium on Code Generation and Optimization* (ACM, Vienna Austria, 2018), pp. 113–125. ISBN: 978-1-4503-5617-6. https://doi.org/10.1145/3168822

19. W. Hattori, S. Yamashita, Quantum circuit optimization by changing the gate order for 2D nearest neighbor architectures, in *International Conference of Reversible Computation*, vol. 11106 (2018), pp. 228–243. https://doi.org/10.1007/978-3-319-99498-7_16

20. Z. Sasanian, D.M. Miller, Reversible and quantum circuit optimization: a functional approach, in *International Conference of Reversible Computation*, ed. by R. Glück, T. Yokoyama, red. by D. Hutchison, T. Kanade, J. Kittler, et al., vol. 7581 (2013), pp. 112–124. https://doi.org/10.1007/978-3-642-36315-3_9

21. G. Vidal, C.M. Dawson, Universal quantum circuit for two-qubit transformations with three controlled-NOT gates. Phys. Rev. A **69**(1), 010–301 (2004). ISSN: 1050-2947, 1094-1622. https://doi.org/10.1103/PhysRevA.69.010301

22. T. Itoko, R. Raymond, T. Imamichi, A. Matsuo, Optimization of quantum circuit mapping using gate transformation and commutation. Integration **70**, 43–50 (2020). ISSN: 0167-9260. https://doi.org/10.1016/j.vlsi.2019.10.004

23. Y. Nam, N. J. Ross, Y. Su, A. M. Childs, D. Maslov, Automated optimization of large quantum circuits with continuous parameters. NPJ Quant. Inf. (2018)

24. D. Maslov, G. Dueck, D. Miller, C. Negrevergne, Quantum circuit simplification and level compaction. IEEE Trans. CAD Integr. Circ. Syst. **27**(3), 436–444 (2008). ISSN: 0278-0070, 1937-4151. https://doi.org/10.1109/TCAD.2007.911334

25. A.K. Prasad, V.V. Shende, I.L. Markov, J.P. Hayes, K.N. Patel, Data structures and algorithms for simplifying reversible circuits. J. Emerg. Technol. Comput. Syst. **2**(4), 277–293 (2006)

26. K. Iwama, Y. Kambayashi, S. Yamashita, Transformation rules for designing CNOT-Based quantum circuits. Des. Autom. Conf. 419–424 (2002)

27. R. Duncan, A. Kissinger, S. Perdrix, J. van de Wetering, Graph-theoretic simplification of quantum circuits with the ZX-calculus. arXiv: 1902.03178 (2019), preprint

28. D. Bhattacharjee, A.A. Saki, M. Alam, A. Chattopadhyay, S. Ghosh, MUQUT: multi-constraint quantum circuit mapping on NISQ computers: invited paper. Int. Conf. CAD (2019)

In the previous chapter, an overview of the compilation flow for quantum circuits was given. In this chapter, we will delve into one of the most important aspects of quantum circuit compilation: quantum circuit mapping. Existing solutions to the mapping problem are mainly heuristic in nature, which is reasonable for an NP-complete problem [1, 2]. However, heuristic approaches raise concerns about the quality of the proposed solutions. It remains unclear how close the results obtained by a heuristic are to the optimum and, thus, how effective the proposed methods are. Therefore, optimal methods are necessary to evaluate the state of the art, even if they can only handle small instances.

In this chapter (based on [3]), an optimal quantum circuit mapping technique is presented based on a symbolic formulation that encodes the whole search space of the mapping problem. To tackle the inherent complexity, sophisticated reasoning engines, such as Boolean satisfiability solvers, are used that are capable of handling huge search spaces. Additionally, performance optimizations are proposed that can generate solutions that are not always minimal but can greatly reduce the solving time while still producing near-optimal results. Experimental evaluations have shown that at time [3] was published, IBM Qiskit's heuristic solution exceeded the lower bound by more than 100 % on average.

The remainder of this chapter is structured as follows: Sect. 10.1 illustrates the general idea of symbolically encoding the problem and determining optimal solutions. Based on that, Sect. 10.2 presents several performance improvements that help reduce the time to solution. Details on the implementation and a discussion of the experimental results are provided in 10.3.

© The Author(s), under exclusive license to Springer Nature Switzerland AG 2026   89
L. Burgholzer and R. Wille, *Design Automation Tools and Software for Quantum Computing*, https://doi.org/10.1007/978-3-032-06770-8_10

10.1   General Idea

In this section, we describe how to determine an optimal solution to the quantum circuit mapping problem, that is, a solution that induces minimal overhead in terms of the number of gates added to the circuit. To this end, we first describe the main idea for tackling the underlying complexity of the problem and, afterward, introduce a symbolic formulation describing the problem in terms of a Boolean function. Eventually, this formulation will be used to optimally solve the problem by applying powerful and efficient reasoning engines.

10.1.1 Main Idea for Tackling the Complexity

Recall that the mapping problem describes the task of determining a representation of an n-qubit quantum circuit $G = g_0, \ldots, g_{|G|-1}$ that conforms to all constraints imposed by an m-qubit architecture (described by a coupling graph (V, E))—all while keeping the overhead of mapping the circuit as small as possible. In the remainder of this chapter, we assume that G has already been decomposed to native gates, and we will mainly focus on minimizing the overhead in terms of the number of gates added during the mapping. This scenario allows G to be restricted to consist only of two-qubit gates, since single-qubit gates are not affected by the limited connectivity and, thus, do not require mapping. Furthermore, we do not consider pre- or post-mapping optimizations (as, e.g., proposed in [4, 5]) that may be applied before or after the mapping, but solely consider the actual mapping process.

As discussed in Sect. 9.1, the mapping problem is generally solved by finding an initial layout for the qubits of the circuit on the device and dynamically inserting *SWAP* gates throughout the mapping process so that any two-qubit gate of the circuit acts only on the qubits connected on the device. To determine an optimal or at least close-to-optimal solution, heuristically or incompletely considering the search space is insufficient. Instead, *all* possible applications of *SWAP* gates that may influence the realization of a given circuit on a certain device.[1] Obviously, this results in a computationally very expensive task. In order to cope with this complexity, powerful reasoning engines, such as solvers for Boolean satisfiability, can be employed (cf. Sect. 3.4).

In this chapter, this reasoning power is used to consider the whole search space of the mapping problem and, by this, to tackle the complexity. To this end, however, a symbolic formulation is required, which completely describes the problem and is provided in terms of a Boolean function (so that it can be used by those reasoning engines).

---

[1] It will be shown later, in Chap. 11, that the search space can be limited while still preserving optimality by carefully analyzing which *SWAP* gates really influence the mapping.

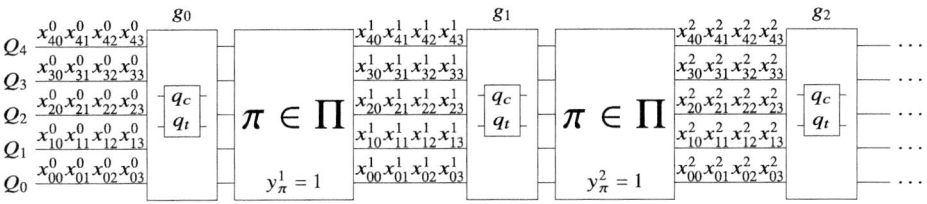

Fig. 10.1 Symbolic formulation for the mapping problem

10.1.2 Symbolic Formulation of the Problem

In the following, a symbolic formulation of the considered problem—quantum circuit mapping—is proposed. To this end, variables are defined that describe all possible applications of the *SWAP* operations. More precisely, those operations basically affect how (logical) qubits from an originally given quantum circuit are mapped to the (physical) qubits of a targeted device. The mapping might be changed before each gate. This leads to the following symbolic formulation:

Let $G = g_0 g_1 \cdots g_k \cdots g_{|G|-1}$ be a quantum circuit composed of $|G|$ two-qubit gates. Each gate g_k operates on two qubits q_c and q_t.[2] Furthermore, let $Q = \{q_0, \ldots, q_{n-1}\}$ denote the (logical) qubits of the circuit that will be mapped to the $m \geq n$ (physical) qubits $V = \{Q_0, \ldots, Q_{m-1}\}$ of the targeted device. Finally, let $E \subseteq V \times V$ be the coupling map describing the connectivity of the device. Then, *mapping variables* x_{ij}^k with $k \in \{0, \ldots, |G| - 1\}, i \in \{0, \ldots, m - 1\}$, and $j \in \{0, \ldots, n - 1\}$ are introduced representing whether, before gate $g_k \in G$, the logical qubit q_j is mapped to the physical qubit Q_i ($x_{ij}^k = 1$) or not ($x_{ij}^k = 0$).

Example 10.1 Consider an arbitrary four-qubit circuit only consisting of two-qubit gates that shall be mapped to a five-qubit architecture. Then, Fig. 10.1 sketches a symbolic formulation for mapping the circuit that represents all possible assignments of the circuit's qubits to the qubits of the architecture. To this end, the leftmost part of Fig. 10.1 represents the initial layout. For example, setting $x_{02}^0 = 1$ implies that q_2 of the circuit is mapped to Q_0 on the architecture right before gate g_0.

Passing this symbolic formulation to a reasoning engine would yield arbitrary assignments of the variables, which would most likely encode impossible or useless mappings (e.g., mapping several circuit qubits to the same device qubit). Hence, *constraints* have to

---

[2] We use the indices c and t because most common two-qubit gates are controlled gates, where one qubit acts as the *control* and one acts as the *target*. Although not all two-qubit gates are controlled gates, the order of the arguments matters for most two-qubit gates, so it makes sense to assign some meaning to it.

be added to the formulation so that only valid solutions are obtained. In that regard, it has to be ensured that

1. A well-defined mapping between circuit qubits and device qubits is conducted, i.e., each qubit of the circuit is uniquely assigned to *exactly one* device qubit and each device qubit is assigned to *at most one* circuit qubit. This is ensured by

$$\bigwedge_{k=0}^{|G|-1} \left(\bigwedge_{j=0}^{n-1} \left(\sum_{i=0}^{m-1} x_{ij}^k = 1 \right) \wedge \bigwedge_{i=0}^{m-1} \left(\sum_{j=0}^{n-1} x_{ij}^k \le 1 \right) \right). \tag{10.1}$$

2. All (two-qubit) gates of the circuit are mapped to the device so that the qubits they act on are connected on the device. This is ensured by

$$\bigwedge_{g_k(q_c, q_t) \in G} \left(\bigvee_{(Q_i, Q_j) \in E} (x_{ic}^k \wedge x_{jt}^k) \right). \tag{10.2}$$

Adding these restrictions and passing the resulting symbolic formulation to a reasoning engine eventually yields a valid solution. Moreover, the resulting formulation covers the entire search space in a symbolic way. Having this, all that is left is a proper description of the costs of the respectively chosen mapping, i.e., the number of *SWAP* gates added by changing the mapping of the qubits throughout the circuit. To properly describe this within the symbolic formulation, the following variables are introduced:

Let $0 \le k \le |G| - 1$ be the index of the gate g_k in a quantum circuit, m the number of qubits in the considered quantum device and $\pi \in \Pi$ a permutation of m elements that indicates how the state of the qubits is permuted (eventually realized by inserting SWAP operations). Then, the *permutation variables* y_π^k indicate whether the permutation π is applied before the gate g_k ($y_\pi^k = 1$) or not ($y_\pi^k = 0$).

Example 10.2 Consider again the symbolic formulation shown in Fig. 10.1. Points in the circuit where the mapping may change are sketched by boxes labeled $\pi \in \Pi$. Here, the assignment of the variable before and after (i.e., the assignments of x_{ij}^{k-1} and x_{ij}^k) may change according to a permutation $\pi \in \Pi$ (eventually to be represented by y_π^k).

These newly introduced variables y_π^k can be linked to the x_{ij}^k variables by introducing

$$\bigwedge_{k=1}^{|G|-1} \left(\bigwedge_{\pi \in \Pi} y_\pi^k \Rightarrow \left(\bigwedge_{i=0}^{m-1} \bigwedge_{j=0}^{n-1} \left(x_{ij}^{k-1} = x_{\pi(i)j}^k \right) \right) \right) \quad \text{and} \quad \bigwedge_{k=1}^{|G|-1} \left(\sum_{\pi \in \Pi} y_\pi^k = 1 \right).$$

$$\tag{10.3}$$

In fact, this ensures that y_π^k is set to 1 iff the assignment of the variables x_{ij}^{k-1} and x_{ij}^k actually describes a change of the mapping defined by π. This cumbersome construction of an implication *plus* a cardinality constraint is necessary since, if $n < m - 1$, π cannot be uniquely determined from the x_{ij}^k variables alone. In case $m - 1 \le n \le m$, the second constraint can be dropped and the implication in the first constraint can simply be replaced by an equivalence.

Satisfying all of the constraints introduced above yields a valid mapping of the originally given circuit to the desired architecture while, at the same time, the costs are determined by

$$\mathcal{F} = \sum_{k=1}^{|G|-1} \sum_{\pi \in \Pi} (\text{swaps}(\pi)\, y_\pi^k). \tag{10.4}$$

Here, $swaps(\pi)$ defines the number of *SWAP* operations needed to realize the permutation π. This value has to be determined for each permutation $\pi \in \Pi$, e.g., by starting with the identity permutation and performing a breadth-first search over single *SWAP* applications until the target permutation is reached. Note that, in theory, this only has to be computed once per architecture and can be reused thereafter. However, in practice, this is only feasible for moderately sized architectures, since it requires the storage of $m!$ cost values for an m-qubit architecture. Assuming the usage of 64 bit integers, a database for a 12-qubit device would require roughly 3.5 GiB, while the database for a 16-qubit device would already require ≈ 152TiB.

Passing the resulting symbolic formulation to a reasoning engine eventually allows one to determine a valid mapping together with the associated cost (i.e., the number of additionally required operations). Since we are also interested in the minimum costs, the cost function $\mathcal{F}$ needs to be further restricted. One direct solution could be to simply set $\mathcal{F}$ to a fixed value and approach towards the minimum, e.g., by applying a binary search. However, since many reasoning engines additionally allow to consider an objective function, the most efficient way is to simply add the objective min: $\mathcal{F}$ to the resulting instance—enforcing the reasoning engine not only to determine a satisfying assignment (representing a valid mapping) but, at the same time, also to minimize $\mathcal{F}$.

10.2 Performance Improvements

Clearly, determining minimal solutions is the desired way to go. However, even with powerful reasoning engines, we cannot always escape the NP-complete complexity of the problem. In this regard, the methodology proposed in the previous section allows for several performance improvements. In fact, the reasoning engine only needs to determine a "minimal" assignment for the x_{ij}^k-variables (the variables y_π^k can be ignored, since they are only used to formulate the costs and their assignments can be directly deduced from the x_{ij}^k-variables). For an n-qubit quantum circuit G with $|G|$ two-qubit gates and an m-qubit architecture, this leads to a total of $n \cdot m \cdot |G|$ Boolean variables to be assigned and therefore an overall search space of $2^{n \cdot m \cdot |G|}$. This search space can easily be restricted by adding further constraints to the x_{ij}^k-variables. Although this may lead to solutions that are no longer guaranteed to be minimal, it can significantly speed up the solving time while, at the same time, remaining very close to minimal (as also confirmed by the experimental evaluations summarized in Sect. 10.3. This section shows possible improvements in this regard. Furthermore, it shows that the quality of the mapping can be improved for directional architectures by adding additional variables and constraints.

10.2.1 Considering Sub-architectures

A scenario frequently occurs where the number of qubits of a given quantum circuit to be mapped is smaller than the number of qubits provided by the architecture (i.e., where $n < m$). Then, obviously, not all physical qubits are required. In fact, this allows to consider only a subset of n device qubits while ignoring the remaining $m - n$ ones. Since the number of device qubits to consider contributes to the search space exponentially, restricting this number yields substantial simplifications. In order to remain as close as possible to the minimal solution, one can try out all $\binom{m}{n}$ possible subsets of qubits (or *sub-architectures*) to consider and solve the resulting (smaller) instances separately. This reduces the overall search space to $\binom{m}{n}2^{n^2 \cdot |G|}$. Note that it has been shown in [6] that this is not guaranteed to preserve optimality in general.

Example 10.3 Consider again the symbolic instance for mapping a four-qubit quantum circuit to a five-qubit architecture sketched in Fig. 10.1. By considering only four device qubits in the mapping procedure, the overall search space for a single instance reduces from $2^{4 \cdot 5 \cdot 5} = 2^{100}$ to $2^{4^2 \cdot 5} = 2^{80}$. Even if all $\binom{5}{4} = 5$ possible subsets of device qubits are considered separately, this still yields a significant reduction in the overall search space.

The search space can be reduced further by only considering sub-architectures that are connected (this can be checked in $O(n)$ time) and possibly removing isomorphic sub-architectures.

Example 10.4 Assume that the circuit for four qubits considered so far shall be mapped to the five-qubit, T-shaped *ibmq_quito* device shown previously in Fig. 9.1c. Then, all valid subsets of device qubits must contain Q_1, since no connected subgraph composed of four nodes without Q_1 is possible. This reduces the number of instances that are passed to the reasoning engine from five $(\binom{5}{4})$ to three—two four-qubit lines and one T-shaped architecture. The removal of the isomorphic sub-architecture reduces the number of instances to two.

See [6] for an even more detailed investigation of the subsets of qubits that should be considered for optimal quantum circuit mapping.

10.2.2 Layering Strategies

So far, we allowed permutations $\pi \in \Pi$ of the mapping before each gate (except the first where an arbitrary initialization can be chosen). While this guarantees minimality (since all possible solutions are considered), this substantially contributes to the complexity. However, in many cases, valid and inexpensive mappings are still possible if permutations of mappings are allowed not before *all* gates $g \in G$, but only before a subset $G' \subseteq G \setminus \{g_0\}$ of them. With $|G'|$ being significantly smaller than G, this reduces the overall search space to $2^{n \cdot m \cdot (|G'|+1)}$.

Upon applying this idea, G' can be chosen arbitrarily. A smaller G' leads to a greater performance improvement, but also to a more restricted instance (yielding solutions that might be far from minimal or even instances for which no valid mapping can be determined anymore). In this work, the following ("layering") strategies for defining G' are considered:

- *Disjoint qubits*, which exploits the fact that gates acting on disjoint sets of qubits can always be mapped in such a way that no intermediate permutations are required. Note that such a set of gates is called a *layer* in some heuristic solutions [4, 7]. To this end, the quantum circuit is clustered into sequences of gates acting on disjoint sets of qubits, and permutations are only allowed before each of those sequences. This corresponds to an as-soon-as-possible (ASAP) scheduling.
- *Odd gates*, which allow permutations only before gates with an odd index. Here, it is still guaranteed that a valid mapping can be determined since either (1) the gates operate on disjoint sets of qubits as discussed above, (2) the gates share both qubits, or (3) the gates share one qubit (and there exists at least one qubit that can interact with two other qubits).
- *Qubit triangle*, which exploits the structure of architectures whose coupling map forms "triangles" of qubits (such as older IBM devices). Here, we can cluster the circuit into sequences of gates, where each sequence acts on at most three qubits. Then, each such sequence of gates can be mapped to a triangle as described above, and permutations are only required before each of those sequences. This concept can be generalized to any fully connected sub-architecture of the overall device.

10.2.3 Improved Mappings on Directed Architectures

Although most architectures nowadays are bidirectional, it is still worth considering direc-
tional architectures in quantum circuit mapping, as outlined in Sect. 9.1. On a directed
architecture, two-qubit gates may only be applied to two (connected) qubits in a certain
direction, that is, if the architecture, for example, supports $CNOT$ gates and contains an
edge (Q_c, Q_t), then a $CNOT(Q_c, Q_t)$ gate may be applied, while a $CNOT(Q_t, Q_c)$ gate
may not. To make $CNOT(Q_t, Q_c)$ applicable, a $SWAP(Q_c, Q_t)$ could be introduced before
the gate.

Example 10.5 On a directed architecture, a $SWAP$ gate is commonly realized as

$$\text{(10.5)}$$

which implies that every $SWAP$ gates incurs seven additional gates (assuming that the
Hadamard gate is native to the device).

However, as the following example illustrates, there might be cheaper ways to fix the
directionality of a two-qubit gate.

Example 10.6 It holds that

$$\text{(10.6)}$$

Thus, the directionality of a $CNOT$ gate could be reversed by inserting pairs of Hadamard
gates before and after the gate. As a result, only four additional gates are necessary compared
to the seven gates required by a $SWAP$ gate.

The additional degree of freedom introduced by this *direction reversal* can be incorporated into the symbolic formulation proposed in this chapter as follows:

- The coupling map constraints are relaxed to permit both directions of an edge in the coupling map, i.e.,

$$\bigwedge_{g_k(q_c.q_t)\in G}\left(\bigvee_{(Q_i.Q_j)\in E}(x_{ic}^k \wedge x_{jt}^k) \vee (x_{it}^k \wedge x_{jc}^k)\right). \tag{10.7}$$

- Variables z^k are introduced for every two-qubit gate $g_k \in G$ that supports a cheap version of direction reversal, i.e., where there is a way to change the direction of the gate by applying a sequence of gates that is cheaper than a *SWAP* gate on the device. These variables are constrained by

$$z^k \Leftrightarrow \bigwedge_{g_k(q_c.q_t)\in G}\left(\bigvee_{(Q_i.Q_j)\in E}(x_{it}^k \wedge x_{jc}^k)\right). \tag{10.8}$$

- The cost function is updated to

$$\mathcal{F} = \sum_{k=1}^{|G|-1}\sum_{\pi\in\Pi}(c_{\text{SWAP}} \cdot \text{swaps}(\pi)\, y_{\pi}^k) + \sum_{k=0}^{|G|-1}(c_{\text{rev}}(g_k) \cdot z^k), \tag{10.9}$$

where c_{SWAP} denoted the cost of a *SWAP* on the device and $c_{\text{rev}}(g_k)$ denotes the cost of reversing the direction of g_k.

In this way, the reasoning engines get to decide which combination of *SWAPs* and direction reversals yields the cheapest global mapping. As demonstrated by experimental evaluations, which are summarized below, this can yield substantially cheaper mappings.

10.3 Summary of Results

The proposed method for optimal quantum circuit mapping has been implemented and serves as a core pillar of the open-source quantum circuit compilation tool QMAP [8] as part of the *Munich Quantum Toolkit* (MQT, [9]). QMAP is open source and publicly available at https://github.com/munich-quantum-toolkit/qmap.

Its core is written in C++ and also provides the modernized reference implementation for the A^* heuristic mapper proposed in [7]. To make the tool as user-friendly as possible, it includes easy-to-use Python bindings, is offered as pre-built Python wheels for all major platforms, and interfaces easily with IBM's Qiskit. As reasoning engine, Microsoft's SMT solver *Z3* [10] is used.

All strategies proposed in this chapter have been extensively evaluated in [3], which led to the following results:

- Determining an optimal mapping is quite expensive—only solutions for instances with rather low gate counts could be determined, which is hardly surprising given that the underlying problem is NP-complete.
- Considering sub-architectures (cf. Sect. 10.2.1) allows to drastically reduce the runtime while still preserving optimality in all considered cases. Note that [6] has shown that this preservation of optimality is not guaranteed in general.
- Clustering the circuit into layers (cf. Sect. 10.2.2) has a tremendous effect on runtime. In fact, the runtime required to solve an instance is directly correlated with $|G'|$, i.e., the lower the number of layers, the faster the time to solution. However, grouping too many gates generates rather poor results regarding optimality.
- A comparison to the heuristic algorithm offered by IBM's Qiskit [4] showed that, at the time [3] was published, Qiskit's solution was producing circuits with an overhead of 104 % on average, that is, the circuits mapped by Qiskit contained twice as many additional gates as necessary.

Overall, even though the optimal mapping approach proposed in this chapter is only applicable for mapping small quantum circuits on small quantum architectures, it shows that there is much room for improvement of heuristic approaches—further motivating research on this topic.

Implementation, Usage, Documentation, and Results

 The proposed optimal mapping technique is available as part of the open-source MQT QMAP tool [8] at https://github.com/munich-quantum-toolkit/qmap, which can be installed using pip install mqt . qmap.

 Using QMAP to map a Qiskit quantum circuit to a given Qiskit backend and get back the mapped Qiskit circuit is as easy as:

```
from mqt import qmap
from qiskit import QuantumCircuit
from qiskit.providers.fake_provider import
    GenericBackendV2

circ = QuantumCircuit(3)
circ.h(0)
circ.cx(0, 1)
circ.cx(0, 2)
circ.measure_all()

arch = GenericBackendV2(
    num_qubits=5,
    coupling_map=[[0, 1], [1, 0], [1, 2], [2, 1],
[1, 3], [3, 1], [3, 4], [4, 3]],)

circ_mapped, results = qmap.compile
        (circ=circ, arch=arch, method='exact')
```

The use _subsets option can be used to configure whether the algorithm considers all possible subarchitectures or the full architecture.

 Documentation on all available configuration options is available at https://mqt . readthedocs . io/projects/qmap

 Details on the experimental setup, evaluations, and results can be found in [3].

References

1. A. Botea, A. Kishimoto, R. Marinescu, On the complexity of quantum circuit compilation, in *International Symposium on Combinatorial Search* (2018)
2. M.Y. Siraichi, V.F. dos Santos, C. Collange, F.M.Q. Pereira, Qubit allocation, in *International Symposium on Code Generation and Optimization* (Vienna Austria, ACM, 2018), pp. 113–125, ISBN: 978-1-4503-5617-6. https://doi.org/10.1145/3168822
3. R. Wille, L. Burgholzer, A. Zulehner, Mapping quantum circuits to IBM QX architectures using the minimal number of SWAP and H operations, in *Design Automation Conference* (2019). https://doi.org/10.1145/3316781.3317859
4. A. Javadi-Abhari, M. Treinish, K. Krsulich, et al., *Quantum Computing with Qiskit* (2024). https://doi.org/10.48550/arXiv.2405.08810. arXiv: 2405.08810 [quant-ph]
5. A. Zulehner, R. Wille, Compiling SU(4) quantum circuits to IBM QX architectures, in *Asia and South Pacific Design Automation Conference* (2019), pp. 185–190. https://doi.org/10.1145/3287624.3287704
6. T. Peham, L. Burgholzer, R. Wille, On optimal sub-architectures for quantum circuit mapping. ACM Trans. Quantum Comput. (2023). arXiv:2210.09321 [quant-ph]. http://arxiv.org/abs/2210.09321
7. A. Zulehner, A. Paler, R. Wille, An efficient methodology for mapping quantum circuits to the IBM QX architectures. IEEE Trans. CAD Integr. Circ. Syst. (2019). https://doi.org/10.1109/TCAD.2018.2846658
8. R. Wille, L. Burgholzer, MQT QMAP: efficient quantum circuit mapping, in *International Symposium on Physical Design* (2023). https://doi.org/10.1145/3569052.3578928
9. R. Wille, L. Berent, T. Forster, et al., The MQT handbook: a summary of design automation tools and software for quantum computing, in *IEEE International Conference on Quantum Software (QSW)* (2024). https://doi.org/10.1109/QSW62656.2024.00013. arXiv: 2405.17543. A live version of this document is available at https://mqt.readthedocs.io
10. L. de Moura, N. Bjørner, Z3: an efficient SMT solver, in *Tools and Algorithms for the Construction and Analysis of Systems*, ed. by C.R. Ramakrishnan, J. Rehof (Springer, 2008), pp. 337–340

Many existing solutions for the quantum circuit mapping problem trade off accuracy or optimality for runtime efficiency [1–9]. The main reason why this trade-off is needed to achieve scalable solutions is that the search space that needs to be considered when determining the best possible mappings increases exponentially with respect to the number of involved qubits. And while heuristic methods are constantly improving, it has been clearly shown that there is a lot of room for improvement, as they frequently stray far from the optimum (see previous chapter). Consequently, several approaches have been proposed to generate *optimal* circuit mappings [10–15]. However, these methods extensively explore the immense search space of the mapping problem—significantly limiting their efficiency and applicability.

In this chapter (based on [16]), we show that this search space can be drastically limited, *while still guaranteeing optimal results*. More precisely, we present generic, architecture-independent observations that allow one to substantially reduce the number of permutations to be considered in front of each gate—the origin of the huge search space and complexity. The key idea is that it suffices to permute just enough that any two qubits of the architecture may interact with each other. Those observations are additionally backed by theoretical considerations (based on group theory), showing that corresponding limitations of the search space are indeed guaranteed to preserve optimality. Based on that, strategies are proposed to utilize these findings in existing approaches for optimal quantum circuit mapping.

Experimental evaluations confirm the resulting benefits. By limiting the search space using the strategies proposed in this chapter, instances that previously suffered from timeouts can now be mapped within minutes or speed-ups of up to three orders of magnitude can be achieved—*all while preserving optimality*.

The remainder of this chapter is structured as follows: In Sect. 11.1, we briefly review the mapping problem and what constitutes an optimal solution to this problem. Then, Sect. 11.2 describes our observations, which allow one to reduce the number of permutations to be considered in front of every gate. Based on that, we then back those observations with a theoretical consideration in Sect. 11.3. In Sect. 11.4, we then propose strategies how these findings can be utilized in existing methods. Details on the implementation and a discussion of the experimental results are provided in Sect. 11.5

11.1 Optimal Solutions for the Mapping Problem

The complexity of the mapping task mainly comes from the fact that, in principle, any possible permutation of the circuit's qubits might be applied in front of each gate of the circuit in order to determine a conforming mapping. Any permutation is eventually realized as a series of architecture-conforming *SWAP* operations. An optimal solution to the mapping problem can be determined by finding the right permutations to apply in front of every gate so that the overall resulting number of *SWAP* gates is minimal. As a result, for a circuit G with $|G|$ gates to be mapped to an m-qubit architecture the search space comprises a total of $|G| * m!$ permutations. Note that, as shown in [15], grouping gates into layers, for example, to capture parallel execution of gates, is not guaranteed to produce gate-optimal results. Therefore, it is necessary to consider an arbitrary permutation in front of every single gate to achieve gate-optimal results.

Example 11.1 Consider the four-qubit quantum circuit G composed of four *CNOT* gates as shown in Fig. 11.1a and assume that it shall be mapped to a four-qubit (linear) architecture described by the coupling graph

$$(V, E) = (\{Q_0, Q_1, Q_2, Q_3\}, \{e_{01}, e_{12}, e_{23}\}), \tag{11.1}$$

which is shown in Fig. 11.1b. Then, Fig. 11.1c shows one possible mapping of G to this architecture. By assigning

$$q_0 \mapsto Q_2, \quad q_1 \mapsto Q_0, \quad q_2 \mapsto Q_3, \quad \text{and} \quad q_3 \mapsto Q_1, \tag{11.2}$$

only a single *SWAP* operation applied to Q_0 and Q_1 is needed in order for all gates to be executable.

Now, let Π denote the set of all permutations of four elements. Then, Fig. 11.1d sketches a symbolic formulation for the mapping task (cf. Fig. 10.1). Conceptually, any permutation $\pi \in \Pi$ can be applied in front of every gate of the circuit. This amounts to $|G| * m! = 4 * 4! = 96$ permutations to be considered for determining an optimal solution.

One of the first methods for tackling the resulting complexity based on a formulation as a satisfiability problem has been proposed in [15], which forms the basis of chapter 10. In parallel and over the following years, several complementary solutions have been proposed. Siraichi et al. used dynamic programming to determine an optimal solution [10], while de Almeida et al. formulated the mapping task as an integer linear programming problem [13].

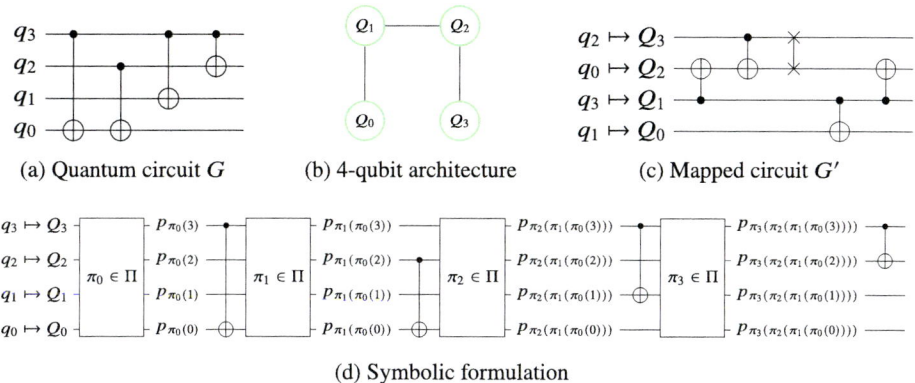

(a) Quantum circuit G (b) 4-qubit architecture (c) Mapped circuit G'

(d) Symbolic formulation

Fig. 11.1 Symbolic formulation for an exemplary mapping problem

A systematic enumeration and pruning technique was presented in [14]. Furthermore, there are works seeking a *time-optimal* mapping using SAT [11] or guided search [12]. All these methods have in common that they eventually explore huge parts of the immense search space to determine an optimal solution. The question arises of whether the search space can be somehow limited while preserving optimality.

11.2 Limiting the Search Space

In the following, we show that the search space for the mapping problem can be significantly limited, *while still guaranteeing optimal results*. This is motivated by three observations, which are described in this section. On the basis of that, we afterward provide a theoretical argument confirming that these observations indeed preserve optimality (in Sect. 11.3) and propose strategies how these findings can be utilized in the methods proposed before (in Sect. 11.4).

The first observation is based on the maximum length of all pairwise shortest paths between two nodes (i.e., device qubits) in a coupling graph (V, E), i.e., the longest direct connection between two nodes. This length is denoted as K in the following and can be determined in a time proportional to $O(|V|^3)$ using, for example, the Floyd-Warshall algorithm [17]. We observe that it is sufficient to permute just enough so that any two qubits can interact with each other, i.e., instead of all $\pi \in \Pi$ permutations, it is sufficient to only consider permutations that can be realized by at most $K - 1$ *SWAP* operations (a more formal argumentation for that is presented later in Sect. 11.3). The more connected the coupling graph (that is, the smaller K in relation to $|V|$, or the larger $|E|$), the easier it is to reach all other qubits from any single qubit. Consequently, this architecture-dependent limitation is most effective whenever the considered architecture is highly connected.

Example 11.2 Consider the linear 4-qubit architecture shown in Fig. 11.1b. Then, the longest (direct) connection involves Q_0 and Q_3—hence, $K = 3$. Allowing only up to $K - 1 = 2$ *SWAP* operations per permutation reduces the number of permutations to be considered in front of every gate from $4! = 24$ to 9, i.e., by more than half!

In addition to the observation above, another way to reduce the effective permutations that need to be considered when mapping a quantum circuit is already demonstrated in [15]. Assume that a quantum circuit uses fewer qubits than the architecture provides (i.e., $n < m$). Then, instead of considering the entire architecture at once, one can consider the mapping problem on each connected subgraph of the targeted architecture composed of n nodes— leading to substantially smaller problems to solve. While there are at most $\binom{m}{n}$ potential subgraphs, the sparse nature of typical coupling graphs implies that the actual number of instances is much lower—consequently reducing the overall number of permutations. As noted in Sect. 10.2.1, the number of subgraphs considered can be further reduced by only considering non-isomorphic subgraphs. By limiting the search space for all the different problem instances on various subgraphs, the number of permutations is reduced even further.

Example 11.3 Assume that the circuit shown in Fig. 11.1a shall be mapped to a five-qubit linear architecture. Instead of having to consider $5! * 4 = 480$ permutations, there are only $4! * 4 = 96$ permutations per connected subgraph of four vertices. There are only two such subgraphs in the linear architecture. Since these are isomorphic and look exactly like the architecture shown in Fig. 11.1b, merely 96 permutations need to be considered, i.e., a reduction by 80%.

Finally, one can observe that the only *SWAP* gates that are relevant before an operation are those that change the position of either of the operation's qubits. All others can be delayed to a later point in time, as they do not serve to make the gate executable. Thus, any permutation that does not change the position of either of an operation's qubits may be ignored. This observation assumes knowledge of the current mapping before each gate and is not feasible for black-box implementations of the mapping problem such as SAT formulations but is applicable to iterative techniques such as informed search algorithms.

Example 11.4 Assume that a *CNOT* operation between Q_0 and Q_3 in a linear four-qubit architecture (as shown in Fig. 11.1b) will be applied. Then, the permutations corresponding to the identity or the swapping of the middle qubits Q_1 and Q_2 can be ignored because they cannot possibly change the executability of the gate. In general, the larger the architecture, the more permutations can be ignored in front of every gate.

Overall, this shows substantial potential in limiting the search space of the problem and, by this, in improving the efficiency of the corresponding optimal methods.

11.3   Preservation of Optimality When Limiting the Search Space

Although the ideas presented above are merely based on observations, this section provides a formal argument why applying these observations still yields optimal results. To this end, we are going to use concepts from group theory [18], which are briefly revisited first.

11.3.1 Group Theory and Permutation Groups

A group $(G, *)$ is a set of elements G equipped with a binary operation $* : G \times G \to G$ such that

- $\forall a, b, c \in G : (a * b) * c = a * (b * c)$ (*associativity*),
- $\exists e \in G \; \forall g \in G : g * e = e * g = g$ (*identity element*), and
- $\forall g \in G \; \exists g' \in G : g * g' = g' * g = e$ (*inverse element*).

If it is clear from the context, we will omit the $*$ when denoting group operations, that is, we will write $g_0 g_1$ instead of $g_0 * g_1$, and we will denote the inverse of an element $g \in G$ by g^{-1}.

Example 11.5 Consider the set Π of all *permutations* of a finite set of elements $P = \{p_0, \dots, p_{n-1}\}$. Any permutation can be written as a set of cycles where cycles of length one are typically not denoted explicitly, e.g., $(01)(23)$ describes a permutation where $p_0 \leftrightarrow p_1$ and $p_2 \leftrightarrow p_3$, while all other elements remain unchanged. The composition of two permutations π_0 and π_1 is defined as

$$(\pi_1 \circ \pi_0) : P \to P$$
$$p \mapsto \pi_1(\pi_0(p)). \tag{11.3}$$

It can be shown that $(\Pi, \circ)$ forms a group—the so-called *permutation group*. Furthermore, it can be easily shown that the permutation group over a set of size n consists of $n!$ elements.

An important concept in group theory is that of *generating sets* of a group. A subset of elements $S \subseteq G$ of a group $(G, *)$ is called a generating set of the group if all elements of G can be generated by repeatedly applying the $*$ operation with the elements of S.

Example 11.6 Consider the group of permutations on the four-element set $P = \{p_0, p_1, p_2, p_3\}$, which contains $4! = 24$ elements. Then, the set $S = \{(01), (12), (23)\}$ consisting of three nearest-neighbor swaps already generates the whole group, e.g., $(012) = (01) \circ (12)$.

The structure of a group $(G, *)$ with respect to a generating set $S \subseteq G$ can be represented as a directed graph—the *Cayley graph* [18]. To this end, the graph vertices are given by the group elements $g \in G$ and, for every $s \in S$, there is an edge $g \xrightarrow{s} g'$ with $g, g' \in G$, if $s * g = g'$.

Example 11.7 Consider again the permutation group over the set $P = \{p_0, p_1, p_2, p_3\}$ and the set of generators $S = \{(01),\ (12),\ (23)\}$. Then, the corresponding Cayley graph is given by

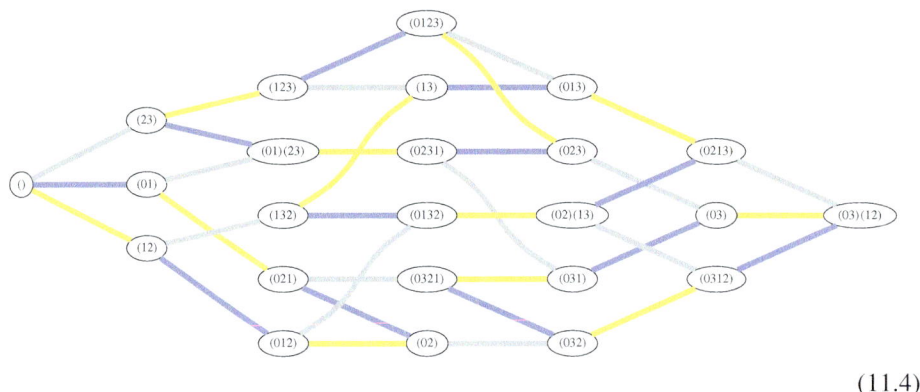

$$(11.4)$$

11.3.2 Theoretical Consideration

In order to understand why the total number of permutations can be reduced drastically while still preserving optimality, we look at the mapping problem from a group theoretic view. Given a coupling graph (V, E), the permutation group over $|V|$ elements can be generated by the set of nearest-neighbor *SWAP* operations executable on the coupling graph, that is, $S = \{(ij) : \forall e_{ij} \in E\}$ generates all permutations of $|V|$ elements for a particular architecture.

Example 11.8 The Cayley graph shown in Example 11.7 illustrates that $S = \{(01), (12), (23)\}$ generates the set of all permutations of four elements. Following any edge with the color corresponding to the generator shows the effect of applying it to the state denoted in the corresponding node. This specific set of generators corresponds to a linear 4-qubit architecture given by $(V, E) = (\{Q_0, Q_1, Q_2, Q_3\}, \{e_{01}, e_{12}, e_{23}\})$ as shown in Fig. 11.1b.

Now, let K be the maximum length of all pairwise shortest paths between two nodes, that is, the longest direct connection between two nodes in the coupling graph. Then, the reduced set of permutations (denoted Π' in the following) is composed of all permutations that can be generated by applying at most $K - 1$ generators from S. This is understood as the construction of the Cayley graph for S, starting with the identity and stopping after $K - 1$ steps.

Example 11.9 Consider again the linear four-qubit architecture shown in Fig. 11.1b with $S = \{(01), (12), (23)\}$. Then, the reduced Cayley graph for this architecture is given by

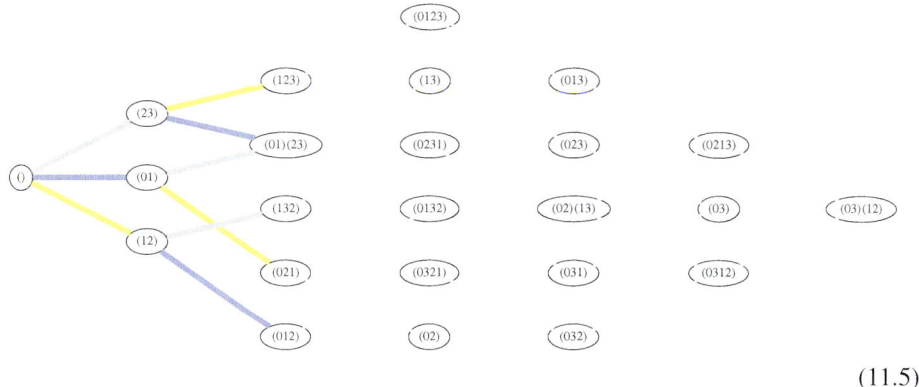

$$(11.5)$$

As already seen in Example 11.2, only 9 permutations need to be considered here (instead of 24).

In order to show that only considering the reduced permutation set Π' preserves optimality, assume otherwise, i.e., assume that at any point in the mapping there exists a permutation $\pi \in \Pi \setminus \Pi'$ that allows for a cheaper overall mapping. More specifically, assume that in front of a $CNOT(q_c, q_t)$ (where q_c is currently mapped to Q_i and q_t to Q_j) there exists $\pi \in \Pi \setminus \Pi'$ such that

1. the gate remains executable (i.e., satisfies the coupling constraints) and
2. the overall amount of *SWAPs* needed to map the circuit is reduced.

Since $\pi \notin \Pi'$, realizing π must at least require K *SWAPs*. Assume that, without loss of generality, $\pi = (kl) \circ \pi'$ for some $\pi' \in \Pi'$. Since K denotes the longest direct path between two qubits, realizing the operation on any other edge of the coupling graph could have already been done using at most $K - 1$ *SWAPs*. Due to the first assumption, any *SWAP* involving i or j that makes the gate non-executable is ruled out. Thus, $k \neq i, j \wedge l \neq i, j$. However, in that case

$$SWAP(Q_k, Q_l)CNOT(Q_i, Q_j) = CNOT(Q_i, Q_j)SWAP(Q_k, Q_l) \qquad (11.6)$$

holds, since gates acting on distinct sets of qubits commute. Consequently, applying the *SWAP* (kl) in front of the current gate cannot reduce the overall cost of the resulting circuit,

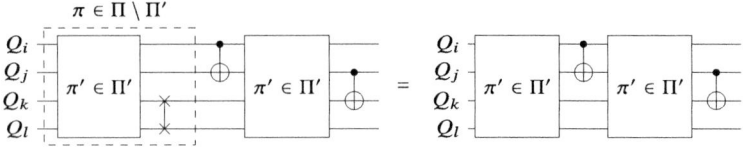

Fig. 11.2 Main circuit identity that allows to show optimality is preserved

since (kl) will be considered once a gate involving either qubit k or l is encountered later on—contradicting the second assumption. Figure 11.2 illustrates this central circuit identity.

This is a strong argument that shows that optimality is preserved if the first observation presented in Sect. 11.2 is applied. A similar argument can be made to argue that the method of ignoring permutations that do not alter the qubits involved in an operation preserves optimality. Although [19] has shown that distributing the problem by considering all possible connected (and non-isomorphic) subgraphs does not necessarily preserve optimality, limiting the permutations considered for each subgraph is guaranteed to preserve the optimal solution for the subproblem following the above argument.

11.4   Resulting Strategies

Based on the considerations from above, the following section proposes two strategies to reduce the number of permutations during quantum circuit mapping. As experiments (summarized in Sect. 11.5) confirm, they allow to substantially reduce the complexity, and hence the runtime of the corresponding optimal mapping methods.

11.4.1 Architecture Limit

In order to capitalize on the first observation from Sect. 11.2, the length K of the longest direct path through the complete architecture has to be determined. It is sufficient to compute this quantity once for a given architecture, for example, by using the Floyd-Warshall algorithm [17], and storing it for future reuse.

Example 11.10 Calculating all pairwise shortest paths for the linear four-qubit architecture shown in Fig. 11.1b results in the following tableau

$$\begin{bmatrix} 0 & 1 & 2 & 3 \\ 1 & 0 & 1 & 2 \\ 2 & 1 & 0 & 1 \\ 3 & 2 & 1 & 0 \end{bmatrix}, \tag{11.7}$$

which allows to determine $K = 3$.

Once K is calculated, the reduced set of permutations Π' to be considered in front of each gate needs to be determined. Since this reduction depends only on the target architecture and is independent of the actual gates to be executed, it can also be computed once, e.g., by constructing a representation of the Cayley graph for the given architecture and stopping after $K - 1$ applications. By defining an ordering of all permutations (e.g., lexicographic ordering), a bitset of size $m!$ may be used to keep track of which permutations are enabled and which are not.

Example 11.11 Calculating the reduced Cayley graph for the linear four-qubit architecture results in

$$\Pi' = \{(), (01), (12), (23), (123), (01)(23), (132), (021), (012)\}, \tag{11.8}$$

as previously shown in Example 11.9. Assuming lexicographic ordering, Π' corresponds to

$$0000\ 0000\ 0001\ 0001\ 1101\ 1111. \tag{11.9}$$

Using Π' instead of Π allows optimal circuit mapping considering a substantially smaller search space. The strategy works best for quantum circuits having as many, or close to as many, qubits as the architecture they get mapped to. This is because this strategy capitalizes on the restrictions imposed due to the inherent structure of the given architecture and limits the number of permutations based on this.

11.4.2 Sub-graph Limit

The second strategy combines dividing the problem into smaller problems on all connected subgraphs with the idea of limiting the number of *SWAP* operations maximally considered in front of each gate. Whenever $n = m$, this naturally becomes the strategy from Sect. 11.4.1. In this fashion, the number of permutations to be considered is further reduced. However, the length K of the longest path now has to be calculated for every possible subgraph of the targeted architecture, instead of the architecture itself. Due to the symmetry/regularity of many quantum architectures, many of these computations are redundant and can be skipped, e.g., by only considering non-isomorphic subgraphs. Subsequently, the subset $\Pi' \subseteq \Pi$ can be determined for every possible connected subgraph in the same way as in Sect. 11.4.1.

Example 11.12 Assume that the circuit shown in Fig. 11.1a is to be mapped to a linear five-qubit architecture. As shown in Example 11.3, there are two connected four-qubit subsets of this architecture—both of which have precisely the structure shown in Fig. 11.1b. Consequently, only a single longest direct path calculation has to be carried out, which results in the same tableau as in Example 11.10—and hence, $K = 3$. The corresponding subsets Π' are characterized by 0000 0000 0001 0001 1101 1111 in both cases.

In general, the benefits of the strategy are greatest whenever $n \ll m$, i.e., whenever a circuit is mapped to a significantly larger architecture. The architecture itself also plays a vital role in the efficiency of this technique. On the one hand, the less connected the architecture is, the fewer connected subsets there are to consider. On the other hand, the less connected the architecture, the larger K (and the larger K, the more permutations must be considered). Consequently, there certainly exists a "sweet spot" of architectural connectivity for using this strategy. As our experimental results (which are summarized in the following) show, the use of this strategy leads to improvements in almost all the cases considered. This shows great promise for applying optimal techniques for the mapping problem to larger circuits and/or larger architectures.

11.5 Summary of Results

The observations and resulting strategies proposed above can, in general, be employed in addition to any optimal method for the mapping problem (such as the one described in Chap. 10 or proposed in [9–15]). As part of the Munich Quantum Toolkit (MQT, [20]), the strategies proposed in Sect. 11.4 have been implemented on top of the open-source MQT QMAP tool [21] introduced in Chap. 10. This led to a version which computes the necessary permutations Π' prior to the execution of the mapping, as well as a version that considers all possible permutations Π. An empirical evaluation of the impact of the proposed search space limitation schemes has been conducted in [16]. The following conclusions can be drawn from these evaluations:

- All considered approaches achieved the optimal mapping cost, independently of whether all permutations Π or just the limited set of permutations Π' have been considered. This is perfectly in line with the theoretical discussion in Sect. 11.3 and confirms that, limiting the search space as proposed in this chapter, still guarantees optimal results.
- Limiting the search space drastically reduces the complexity of the problem. In many cases, it makes the difference between running into a timeout of 1 h or being able to determine an optimal result in just a few minutes. Even in cases where the reference approach manages to obtain a result within 1 h, speed-ups of up to three orders of magnitude can be observed.

Overall, the results demonstrate the drastic impact that the size of the search space has on the execution time of optimal mapping techniques. They also confirm that the methods proposed in this chapter are feasible for limiting the search space while still preserving optimality.

Implementation, Usage, Documentation, and Results

 The proposed search space limitation techniques available as part of the open-source MQT QMAP tool [21] at
https : //github . com/munich-quantum-toolkit/qmap.
which can be installed using pip install mqt. qmap.

 The resulting tool can be set up as described and illustrated in section 10.3 on page 82. Since it is guaranteed that the search space limitation preserves optimality, it is enabled by default when calling

qmap . **compile** (c i r c ,   a r c h ,   method=' e x a c t ')

It can be further configured via the swap_reduction   option.

 Documentation on all available configuration options is available at
https://mqt . readthedocs . io/projects/qmap

 Details on the experimental setup, evaluations, and results can be found in [16].

References

1. A. Zulehner, A. Paler, R. Wille, An efficient methodology for mapping quantum circuits to the IBM QX architectures. IEEE Trans. CAD Integr. Circ. Syst. (2019). https://doi.org/10.1109/TCAD.2018.2846658
2. K.N. Smith, M.A. Thornton, A quantum computational compiler and design tool for technology-specific targets, in *International Symposium on Computer Architecture* (2019), pp. 579–588
3. G. Li, Y. Ding, Y. Xie, Tackling the qubit mapping problem for NISQ-era quantum devices, in *International Conference on Architectural Support for Programming Languages and Operating Systems* (2019). https://doi.org/10.1145/3297858.3304023
4. A. Matsuo, W. Hattori, S. Yamashita, Reducing the overhead of mapping quantum circuits to IBM Q system, in *IEEE International Symposium on Circuits and Systems* (2019)
5. P. Murali, J.M. Baker, A. Javadi-Abhari, F.T. Chong, M. Martonosi, Noise-adaptive compiler mappings for noisy intermediate-scale quantum computers, in *International Conference on Architectural Support for Programming Languages and Operating Systems* (ACM, Providence RI USA, 2019), pp. 1015–1029. ISBN: 978-1-4503-6240-5. https://doi.org/10.1145/3297858.3304075

6. M. Amy, V. Gheorghiu, Staq—a full-stack quantum processing toolkit. Quant. Sci. Technol. **5**(3), 034–016 (2020)
7. S. Sivarajah, S. Dilkes, A. Cowtan, W. Simmons, A. Edgington, R. Duncan, T|ket⟩: a retargetable compiler for NISQ devices. Quant. Sci. Technol.**6**(1), 014–003 (2021). ISSN: 2058-9565. https://doi.org/10.1088/2058-9565/ab8e92
8. A. Zulehner, R. Wille, Compiling SU(4) quantum circuits to IBM QX architectures, in *Asia and South Pacific Design Automation Conference* (2019), pp. 185–190. https://doi.org/10.1145/3287624.3287704
9. Y. Hirata, M. Nakanishi, S. Yamashita, Y. Nakashima, An efficient conversion of quantum circuits to a linear nearest neighbor architecture. Quant. Inf. Comput. **11**, 142–166 (2011). https://doi.org/10.26421/QIC11.1-2-10
10. M.Y. Siraichi, V.F. dos Santos, C. Collange, F.M.Q. Pereira, Qubit allocation, in *International Symposium on Code Generation and Optimization* (ACM, Vienna Austria, 2018), pp. 113–125, ISBN: 978-1-4503-5617-6. https://doi.org/10.1145/3168822
11. C. Zhang, A.B. Hayes, L. Qiu, Y. Jin, Y. Chen, E.Z. Zhang, Time-optimal qubit mapping, in *International Conference on Architectural Support for Programming Languages and Operating Systems* (2021). https://doi.org/10.1145/3445814.3446706
12. B. Tan, J. Cong, Optimal layout synthesis for quantum computing, in *International Conference on CAD* (2020). https://doi.org/10.1145/3400302.3415620
13. A.A.A. de Almeida, G.W. Dueck, A.C.R. da Silva, Finding optimal qubit permutations for IBM's quantum computer architectures, in *Symposium on Integrated Circuits and Systems Design* (2019)
14. P. Zhu, X. Cheng, Z. Guan, An exact qubit allocation approach for NISQ architectures. Quant. Inf. Process. **19**(11), 391 (2020)
15. R. Wille, L. Burgholzer, A. Zulehner, Mapping quantum circuits to IBM QX architectures using the minimal number of SWAP and H operations, in *Design Automation Conference* (2019). https://doi.org/10.1145/3316781.3317859
16. L. Burgholzer, S. Schneider, R. Wille, Limiting the search space in optimal quantum circuit mapping, in *Asia and South Pacific Design Automation Conference* (2022). https://doi.org/10.1109/ASP-DAC52403.2022.9712555
17. T.H. Cormen, C.E. Leiserson, R.L. Rivest, C. Stein, *Introduction to Algorithms*, 3rd edn. (The MIT Press, 2009)
18. N.C. Carter, *Visual Group Theory* (Mathematical Association of America, 2009)
19. T. Peham, L. Burgholzer, R. Wille, On optimal Sub-architectures for quantum circuit mapping. ACM Trans. Quant. Comput. (2023). arXiv:2210.09321 [quant-ph]. http://arxiv.org/abs/2210.09321
20. R. Wille, L. Berent, T. Forster, et al., The MQT handbook: a summary of design automation tools and software for quantum computing, in *IEEE International Conference on Quantum Software (QSW)* (2024). https://doi.org/10.1109/QSW62656.2024.00013. arXiv:2405.17543. A live version of this document is available at https://mqt.readthedocs.io
21. R. Wille, L. Burgholzer, MQT QMAP: efficient quantum circuit mapping, in *International Symposium on Physical Design* (2023). https://doi.org/10.1145/3569052.3578928

Summary of Part III

This part of the book explored the challenge of determining optimal solutions to one of the most important problems in the compilation of quantum circuits: mapping quantum circuits to architectures that have a limited connectivity between their qubits. This is an incredibly challenging problem to solve optimally due to its immense search space. Nevertheless, the development of optimal solutions is crucial to establish lower bounds on the achievable performance and to evaluate the quality of the results from heuristic methods.

The proposed solutions encode the respective problem in a symbolic fashion and use powerful reasoning engines (in particular Microsoft's SMT solver Z3 [1]) to cope with the vast complexity of the underlying problem. More specifically:

- A method has been proposed that allows to determine optimal solutions for the quantum circuit mapping problem, i.e., circuits with as little overhead in terms of additional gates as possible. The resulting solution was used to demonstrate that established heuristic solutions frequently stray far from the optimum and leave much room for improvement. Moreover, several performance optimizations have been proposed that allow to significantly reduce the required runtime while still yielding close-to-optimal solutions.
- Generic and architecture-independent strategies have been proposed to limit the search space that must be considered when trying to determine the optimal results for the quantum circuit mapping problem. These strategies are motivated by observations showing that only a limited set of permutations in front of each gate needs to be considered— addressing the origin of the huge search space and, as a result, drastically improving the performance of corresponding approaches while, at the same time, remaining optimal.

© The Author(s), under exclusive license to Springer Nature Switzerland AG 2026
L. Burgholzer and R. Wille, *Design Automation Tools and Software for Quantum Computing*, https://doi.org/10.1007/978-3-032-06770-8_12

All efforts listed above have contributed towards the development of the open-source quantum circuit compilation tool MQT QMAP [2], which is available at https://github.com/ munich-quantum-toolkit/qmap. QMAP constitutes a push-button, accessible, state-of-the-art tool for determining optimal results to challenging quantum circuit compilation problems. https://github.com/munich-quantum-toolkit/qmap

References

1. L. de Moura, N. Bjørner, Z3: An efficient SMT solver, in *Tools and Algorithms for the Construction and Analysis of Systems*, ed. by C.R. Ramakrishnan, J. Rehof (Springer, 2008), pp. 337–340
2. R. Wille, L. Burgholzer, MQT QMAP: efficient quantum circuit mapping, in *International Symposium on Physical Design* (2023). https://doi.org/10.1145/3569052.3578928

Compiling quantum algorithms as reviewed in Part III results in different representations of the considered functionality, which significantly differ in their basis operations and structure but are still supposed to be functionally equivalent. Consequently, checking whether the original functionality is indeed maintained throughout all these different abstractions becomes increasingly relevant in order to guarantee a consistent and error-free design flow. This is similar to the classical realm, where descriptions at various levels of abstraction also exist. These descriptions are verified using design automation expertise—resulting in efficient methods for verification to ensure the correctness of the design across different levels of abstraction [1–6]. However, since quantum circuits additionally employ quantum-physical effects such as superposition and entanglement, these methods cannot be used out of the box in the quantum realm. Accordingly, how to conduct the verification of quantum circuits must be approached from a different perspective.

To this end, it is important to acknowledge that different understandings about what verification is supposed to accomplish currently exist (as also confirmed, e.g., in [7]). More precisely, the literature distinguishes, e.g., between

- Equivalence Checking (or logic verification), i.e., checking whether a circuit is logically equivalent or at least "close enough" to another circuit (the problem sketched above),
- Quantum Verification, i.e., checking whether a device indeed performs quantum logic (see [8]) and, therefore, indeed utilizes quantum advantage, or
- Physical Verification, i.e., checking whether the hardware implementation performs as intended (see [9]).

This part of the book focuses on equivalence checking of quantum circuits. To this end, it gives an overview of several contributions towards various aspects of this domain in a cumulative fashion, and, by this, creates the first comprehensive, automated, and scalable methodology for quantum circuit equivalence checking. The precise contributions are as follows: First, a general equivalence checking methodology (based on [10–12]) is proposed that does not take the paradigm of quantum computing as a burden, but explicitly exploits quantum characteristics to efficiently determine whether two quantum circuits are equivalent or not. Then, it is shown how this methodology can be tailored toward the verification of compilation flow results (based on [13]) and how different stimuli generation schemes for simulative verification allow for a trade-off between expressiveness and runtime (based on [14]). To complement the approaches mentioned above, a complementary approach to equivalence checking that uses the ZX-Calculus is also proposed (based on [15, 16]). Then, the resulting core methodology is subsequently extended (based on [17]) to support a larger class of quantum circuits, so-called *dynamic quantum circuits*, which combine quantum operations with classical feedback and hence require special treatment. All of the above approaches are then cleverly combined to make verification of parameterized quantum circuits possible (based on [18]).

To set the stage for the next chapters, Sect. 13.1 gives a brief overview of equivalence checking for classical circuits. After that, Sect. 13.2 introduces the equivalence checking problem for quantum computing as considered in the rest of this book. Ultimately, Sect. 13.3 provides an overview of the relevant work in this field.

13.1   Equivalence Checking of Classical Circuits

To ensure that classical circuits function correctly and as intended, verification techniques are essential. By comparing a given circuit, the *Design Under Verification* (DUV), to an also given *Golden Specification*, these methods are used to demonstrate or prove whether the circuit adheres to its specification. *Simulative verification* [1–6] and *formal verification* [19, 20] are the two main verification techniques used in modern industrial practice.

During simulative verification, input assignments (*stimuli*) are assigned to the circuit and propagated through it. Subsequently, the resulting outputs are compared with the expected values. It is favored in many applications because it is faster and simpler to implement than formal verification. However, its main drawback is that the quality of its results heavily depends on the applied stimuli. While an exhaustive set of stimuli would allow one to prove correctness with certainty, it is intractable in practice due to the exponential number of simulations that would be required. *Constraint-based random simulation* [1–4] and *fuzzing* [5, 6] have been proposed as techniques to address these difficulties. They involve the generation of particular input stimuli (e.g., from constraints, mutations of randomly generated inputs, etc.) that explicitly trigger corner cases or cover a wide range of cases. This increases the likelihood of discovering errors that might otherwise go undetected.

On the other hand, formal verification offers a much higher quality solution and 100% certainty of correctness. Formal verification involves mathematical analysis of the problem and proving that the circuit adheres to its specification. In contrast to simulative verification, formal verification frequently employs more complex methods, which, particularly for larger designs, make it impractical due to the exponential complexity of many of these methods.

Although significant progress has been made in the past, verification of classical circuits remains difficult. This is especially true for large designs, where the complexity of verification scales exponentially. As a result, a *verification gap* has emerged, where we are able to design increasingly complex systems, but we are no longer able to successfully verify their functionality. It is the subject of ongoing research to close or, at least, shrink this gap.[1]

13.2 Equivalence Checking of Quantum Circuits

In the context of quantum computing, equivalence checking involves proving that two quantum circuits, G and G', are functionally equivalent or demonstrating their non-equivalence via a counterexample. This approach is comparable to the verification of classical circuits, with one circuit serving as the specification and the other representing the device being verified. Note that these terms have not yet been fully established in the realm of quantum computing, so the focus of this part of the book will be on verifying or checking the equivalence of two circuits. In general, given two quantum circuits $G = g_0 \ldots g_{|G|-1}$ and $G' = g'_0 \ldots g'_{|G'|-1}$, with corresponding system matrices $U = U_{|G|-1} \cdots U_0$ and $U' = U'_{|G'|-1} \cdots U'_0$, the *equivalence checking problem for quantum circuits* asks whether

$$U = e^{i\theta} U', \tag{13.1}$$

where $\theta \in (-\pi, \pi]$ denotes a physically unobservable global phase. So, in principle, checking the equivalence of two quantum circuits reduces to a comparison of the respective system matrices. Furthermore, if U and U' differ in any column i (by more than a global phase factor $e^{i\theta}$), then the corresponding circuits G and G' are not equivalent and $|i\rangle$ serves as a counterexample showing that

$$U |i\rangle = |u_i\rangle \neq |u'_i\rangle = U' |i\rangle. \tag{13.2}$$

Here, the fidelity $\mathcal{F}$ between two states $|x\rangle$ and $|y\rangle$ is typically used as a similarity measure to compare quantum states, where $\mathcal{F}$ is calculated as the squared overlap between states, that is, $\mathcal{F} = |\langle x|y\rangle|^2 \in [0, 1]$. Two states are considered equivalent if the fidelity between them is 1 (up to a given tolerance ε).

Unfortunately, the entire functionality U (and similarly U') is not readily available for comparison, but must be constructed from the individual gate descriptions g_i, which necessitates the sequence of matrix-matrix multiplications

---

[1] In an interesting twist, the progress made in this book toward quantum circuit equivalence checking sparked some complementary ideas about how to close the verification gap in classical circuits [21].

$$U^{(0)} = U_0, \quad U^{(j)} = U_j \cdot U^{(j-1)} \text{ for } j = 1, \ldots, |G| - 1 \tag{13.3}$$

to build the matrix of the whole system $U = U^{(|G|-1)}$.

Example 13.1 Consider the following three-qubit quantum circuits G and G':

$$\tag{13.4}$$

Consecutively multiplying the corresponding gate matrices in both cases eventually yields

$$U = U' = \frac{1}{\sqrt{2}} \begin{bmatrix} 1 & 0 & 1 & 0 & 0 & 0 & 0 & 0 \\ 0 & 1 & 0 & 1 & 0 & 0 & 0 & 0 \\ 0 & 0 & 0 & 0 & 1 & 0 & -1 & 0 \\ 0 & 0 & 0 & 0 & 0 & 1 & 0 & -1 \\ 0 & 1 & 0 & -1 & 0 & 0 & 0 & 0 \\ 1 & 0 & -1 & 0 & 0 & 0 & 0 & 0 \\ 0 & 0 & 0 & 0 & 1 & 0 & 1 & 0 \\ 0 & 0 & 0 & 0 & 0 & 1 & 0 & 1 \end{bmatrix}. \tag{13.5}$$

Consequently, both circuits G and G' are equivalent.

Consider the same scenario as above, but additionally assume that due to an error, the last T-gate of G' (i.e., g'_{15}) is not applied (yielding a new circuit $\tilde{G}'$). Then, a new functionality results, which is described by the system matrix

$$\tilde{U}' = \frac{1}{\sqrt{2}} \begin{bmatrix} 1 & 0 & 1 & 0 & 0 & 0 & 0 & 0 \\ 0 & 1 & 0 & 1 & 0 & 0 & 0 & 0 \\ 0 & 0 & 0 & 0 & 1 & 0 & -1 & 0 \\ 0 & 0 & 0 & 0 & 0 & 1 & 0 & -1 \\ 0 & -\omega^3 & 0 & \omega^3 & 0 & 0 & 0 & 0 \\ -\omega^3 & 0 & \omega^3 & 0 & 0 & 0 & 0 & 0 \\ 0 & 0 & 0 & 0 & -\omega^3 & 0 & -\omega^3 & 0 \\ 0 & 0 & 0 & 0 & 0 & -\omega^3 & 0 & -\omega^3 \end{bmatrix}, \tag{13.6}$$

where $\omega = \frac{1+i}{\sqrt{2}}$. Since U and $\tilde{U}'$ are obviously not identical anymore, the circuits G and $\tilde{G}'$ have been shown to be non-equivalent. Furthermore, since both matrices differ, for example, in column four, $|100\rangle$ serves as a counterexample showing that

$$U \, |100\rangle = |u_4\rangle = \tfrac{1}{\sqrt{2}}[0\ 0\ 1\ 0\ 0\ 0\ 1\ 0]^\top, \text{ while } \tilde{U}' \, |100\rangle = |\tilde{u}'_4\rangle = \tfrac{1}{\sqrt{2}}[0\ 0\ 1\ 0\ 0\ 0\ \tfrac{1-i}{\sqrt{2}}\ 0]^\top.$$
$$\tag{13.7}$$

It holds that $\mathcal{F}(U \ket{100}, \tilde{U}' \ket{100}) = \mathcal{F}(\ket{u_4}, \ket{\tilde{u}_4'}) \approx 0.92 < 1$.

While equivalence checking of quantum circuits is straightforward conceptually, it quickly becomes an increasingly difficult task because the size of the involved matrices scales exponentially with the number of qubits. In fact, equivalence checking of quantum circuits has been shown to be QMA-complete [22], where QMA is the quantum analogue of the classical complexity class NP and, in fact, includes NP.

13.3   Related Work

Methods for checking the equivalence of quantum circuits that have been proposed in the past broadly fall into three categories:

1. *Fully automated methods capable of checking the equivalence of two specific quantum circuits* [23–29]: While these methods can show the equivalence of many quantum circuits, these existing methodologies still remain unsatisfactory in many cases. The method proposed in [23] only works with a limited gate set, as all verification rules must be manually derived explicitly for each gate and parameter. The approaches proposed in [25–29] work with quantum decision diagrams (cf. Sect. 3.1). Although decision diagrams often allow for representing quantum functionality in a very compact fashion, the decision diagrams corresponding to the circuits G and G' may still grow exponentially in the worst case. Finally, [24] expresses rotations in the form of dyadic fractions (similar to [30]). But dyadic fractions are insufficient when working with gates that are not in the Clifford hierarchy, since arbitrary rotations can only be approximated with dyadic fractions.
2. *Semi-automatic methods using formal languages for quantum computing and proof assistants* [31–33]: While these methods are very general when it comes to proving equality in quantum computing, employing them requires expert knowledge about the languages and tools being used, and therefore, they are not suitable for verification from a design automation perspective.
3. *Compiler- and domain-specific methods* [34–36]: These methods are based on the idea of writing compiler passes in a domain-specific language. These languages are constructed in a way that allows for automated verification of entire compilation flows, and, therefore, are more general than an equivalence checking method that only checks the equivalence of two specific circuits. The drawback of these methods is that they require dedicated implementations of compilation flows, which frequently have to be updated once even small aspects of the compiler are changed. Automated equivalence checking, on the other hand, is agnostic to future improvements in compilation and optimization methods, as it only operates on the circuit level and not the compilation level.

Many of the methods mentioned above only work on small circuits, lack publicly available implementations, or are based on classical computing paradigms that do not reflect the complete picture of quantum computing but rather consider it a burden to be dealt with. As a result of the shortcomings of existing methods, this book proposes novel methods for efficient and automated equivalence checking of quantum circuits. Individual contributions will be summarized in the chapters that follow.

References

1. J. Yuan, C. Pixley, A. Aziz, *Constraint-Based Verification* (Springer, 2006)
2. J. Bergeron, *Writing Testbenches Using System Verilog* (Springer, 2006)
3. N. Kitchen, A. Kuehlmann, Stimulus generation for constrained random simulation, in *International Conference on CAD* (2007), pp. 258–265
4. R. Wille, D. Große, F. Haedicke, R. Drechsler, SMT-based stimuli generation in the SystemC Verification library, in *Forum on Specification and Design Languages* (2009)
5. H.M. Le, D. Große, N. Bruns, R. Drechsler, Detection of hardware Trojans in SystemC HLS designs via coverage-guided fuzzing, in *Design, Automation & Test in Europe Conference* (2019)
6. K. Laeufer, J. Koenig, D. Kim, J. Bachrach, K. Sen, RFUZZ: coverage-directed fuzz testing of RTL on FPGAs, in *International Conference on CAD* (2018)
7. Y. Naveh, E. Kashefi, J.R. Wootton, K. Bertels, Theoretical and practical aspects of verification of quantum computers, in *Design, Automation and Test in Europe* (IEEE, Dresden, 2018), pp. 721–730. ISBN: 978-3-9819263-0-9. https://doi.org/10.23919/DATE.2018.8342103
8. Z. Brakerski, P. Christiano, U. Mahadev, U. Vazirani, T. Vidick, A cryptographic test of quantumness and certifiable randomness from a single quantum device. Found. Comput. Sci. **2018**. https://doi.org/10.1109/focs.2018.00038
9. A. Gheorghiu, T. Kapourniotis, E. Kashefi, Verification of quantum computation: an overview of existing approaches. Theory Comput. Syst. **63**(4), 715–808 (2019). ISSN: 1433-0490. https://doi.org/10.1007/s00224-018-9872-3
10. L. Burgholzer, R. Wille, Advanced equivalence checking for quantum circuits. IEEE Trans. CAD Integr. Circuits Syst. (2021). https://doi.org/10.1109/TCAD.2020.3032630
11. L. Burgholzer, R. Wille, Improved DD-based equivalence checking of quantum circuits, in *Asia and South Pacific Design Automation Conference* (2020)
12. L. Burgholzer, R. Wille, The power of simulation for equivalence checking in quantum computing, in *Design Automation Conference* (2020)
13. L. Burgholzer, R. Raymond, R. Wille, Verifying results of the IBM Qiskit quantum circuit compilation flow, in *International Conference on Quantum Computing and Engineering* (2020). https://doi.org/10.1109/QCE49297.2020.00051
14. L. Burgholzer, R. Kueng, R. Wille, Random stimuli generation for the verification of quantum circuits, in *Asia and South Pacific Design Automation Conference* (2021). https://doi.org/10.1145/3394885.3431590
15. T. Peham, L. Burgholzer, R. Wille, Equivalence checking of quantum circuits with the ZX-Calculus. JETCAS (2022). https://doi.org/10.1109/JETCAS.2022.3202204
16. T. Peham, L. Burgholzer, R. Wille, Equivalence checking paradigms in quantum circuit design: a case study, in *Design Automation Conference* (2022)
17. L. Burgholzer, R. Wille, Handling non-unitaries in quantum circuit equivalence checking, in *Design Automation Conference* (2022). https://doi.org/10.1145/3489517.3530482

18. T. Peham, L. Burgholzer, R. Wille, Equivalence checking of parameterized quantum circuits: verifying the compilation of variational quantum algorithms, in *Asia and South Pacific Design Automation Conference* (2023). https://doi.org/10.1145/3566097.3567932

19. A. Biere, W. Kunz, SAT and ATPG: Boolean engines for formal hardware verification, in *International Conference on CAD* (2002), pp. 782–785

20. R. Drechsler, *Advanced Formal Verification* (Springer, 2004)

21. L. Burgholzer, R. Wille, Exploiting reversible computing for verification: Potential, possible paths, and consequences, in *ASPDAC* (2023)

22. D. Janzing, P. Wocjan, T. Beth, Non-identity check is QMA-complete. Int. J. Quant. Inform. **03**(03), 463–473 (2005)

23. W. Chun-Yu, T. Yuan-Hung, J. Chaio-Shan, J. Jie-Hong, Accurate BDD-based unitary manipulation for scalable and robust quantum circuit verification, in *Design Automation Conference* (2022). https://doi.org/10.1145/3489517.3530481

24. M. Amy, Towards large-scale functional verification of universal quantum circuits. Int. Conf. Quant. Phys. Logic (2019)

25. S.-A. Wang, C.-Y. Lu, I.-M. Tsai, S.-Y. Kuo, An XQDD-based verification method for quantum circuits. IEICE Trans. Fundam. (2008), pp. 584–594. https://doi.org/10.1093/ietfec/e91-a.2.584

26. P. Niemann, R. Wille, D.M. Miller, M.A. Thornton, R. Drechsler, QMDDs: efficient quantum function representation and manipulation. IEEE Trans. CAD Integr. Circ. Syst. (2016)

27. G.F. Viamontes, I.L. Markov, J.P. Hayes, Checking equivalence of quantum circuits and states, in *International Conference on CAD* (2007)

28. P. Niemann, R. Wille, R. Drechsler, Equivalence checking in multi-level quantum systems, in *International Conference of Reversible Computation* (2014)

29. S. Yamashita, I.L. Markov, Fast equivalence-checking for quantum circuits, in *International Symposium on Nanoscale Architectures* (2010). https://doi.org/10.1109/NANOARCH.2010.5510932

30. P. Niemann, A. Zulehner, R. Drechsler, R. Wille, Overcoming the tradeoff between accuracy and compactness in decision diagrams for quantum computation. IEEE Trans. CAD Integr. Circ. Syst. **39**(12), 4657–4668 (2020). ISSN: 1937-4151. https://doi.org/10.1109/TCAD.2020.2977603

31. L. Zhou, N. Yu, M. Ying, An applied quantum Hoare logic, in *Conference on Programming Language Design and Implementation*, ser. PLDI (2019), pp. 1149–1162. ISBN: 978-1-4503-6712-7. https://doi.org/10.1145/3314221.3314584

32. E. D'hondt, P. Panangaden, Quantum weakest preconditions. Math. Struct. Comp. Sci. **16**(3), 429–451 (2006). ISSN: 0960-1295. https://doi.org/10.1017/S0960129506005251

33. M. Lewis, S. Soudjani, P. Zuliani, formal verification of quantum programs: theory, tools and challenges. arXiv:2110.01320 [quant-ph] (2021), preprint

34. R. Tao, Y. Shi, J. Yao, et al., Giallar: push-button verification for the Qiskit quantum compiler, in *International Conference on Programming Language Design and Implementation* (ACM, San Diego CA USA, 2022), pp. 641–656. ISBN: 978-1-4503-9265-5. https://doi.org/10.1145/3519939.3523431

35. L. Li, F. Voichick, K. Hietala, Y. Peng, X. Wu, M. Hicks, Verified compilation of quantum oracles. arXiv:2112.06700 [quant-ph]. (2022), preprint

36. R. Rand, J. Paykin, D.-H. Lee, S. Zdancewic, ReQWIRE: reasoning about reversible quantum circuits. Electron. Proc. Theor. Comput. Sci. **287**, 299–312 (2019). ISSN: 2075-2180. https://doi.org/10.4204/EPTCS.287.17. arXiv:1901.10118 [cs]

Advanced Equivalence Checking for Quantum Circuits

In this chapter (based on [1–3]), an advanced equivalence checking methodology is proposed that explicitly utilizes characteristics unique to quantum computing in order to substantially improve existing approaches. More precisely, we unearth potential which rests on the following two observations:

- Quantum circuits are inherently reversible. Because of this, if two quantum circuits G and G' are equivalent, then concatenating the first circuit G with the inverse G'^{-1} of the second circuit would realize the identity function I, i.e., $G \cdot G'^{-1} = I$. The potential now lies in the order in which the operations of either circuit are applied. Whenever a strategy can be employed so that the respective gates from G and G' are applied in a fashion frequently yielding the identity, the entire procedure can be conducted rather efficiently since the identity constitutes the best case for most representations of quantum functionality (e.g., linear in the number of qubits for decision diagrams).
- Moreover, even in the case where the two considered quantum circuits are not equivalent, quantum characteristics can be exploited. In fact, we observed that, again due to the inherent reversibility of quantum operations, even small differences in quantum circuits frequently affect the *complete* functional representation. Hence, it may not always be necessary to check the complete functionality, but it is highly likely that the simulation of both computations with a couple of arbitrary input states (i.e., considering only a small part of the whole functionality) will already provide a counterexample in case of non-equivalence. This is in stark contrast to the classical realm, where the inevitable information loss introduced by many logic gates and the resulting masking effects often require a complete consideration of *all* possible input states or sophisticated schemes for constraint-based stimuli generation [4–7] or fuzzing [8, 9].

L. Burgholzer and R. Wille, *Design Automation Tools and Software for Quantum Computing*, https://doi.org/10.1007/978-3-032-06770-8_14

Both observations provide the nucleus of an advanced equivalence checking methodology in which the non-equivalence is often detected by a few simulation runs (which can be conducted dramatically faster than the actual equivalence check). Moreover, passing several simulation runs leading to the same results for both circuits provides an indication (albeit no proof) that the circuits might be equivalent. The proof itself can be significantly accelerated afterward by using strategies that keep the check for $G \cdot G'^{-1} = I$ close to the identity I. Combining these complementary ideas into a comprehensive equivalence checking flow allows for efficient verification of quantum circuits—in many cases, just a single simulation run is sufficient. By this, we not only show ways to handle the complexity of verifying quantum circuits but also show the potential of simulation for this task.

The remainder of this chapter is structured as follows: Sect. 14.1 introduces and illustrates the general ideas proposed in this chapter. Based on that, Sect. 14.2 describes the dedicated equivalence checking schemes resulting from these ideas, and Sect. 14.3 provides a theoretical discussion on the power of simulation for equivalence checking of quantum circuits. All these findings eventually result in an advanced equivalence checking methodology, which is described and discussed in Sect. 14.4.

14.1 General Ideas

In this section, we describe the general ideas developed in this work to address the shortcomings of the related work and to solve the equivalence checking problem much faster than ever before—in many cases, with just a single simulation run. To this end, we utilize certain characteristics of quantum computing. While these, at first glance, make the problem of verification harder compared to the classical realm (circuits have to be supported that not only rely on zeros and ones but also on superposition or entanglement), they also offer potential that has not really been exploited yet. In the following, we sketch this unearthed potential before dedicated equivalence checking schemes and flows resulting from these ideas are covered in the remaining sections of this chapter.

14.1.1 Potential Power of Reversibility

Many classical logic operations are irreversible (e.g., $x \wedge y = 0$ does not allow one to determine the precise values of x and y). As there is no bijective mapping between input and output states, in general, the concept of the *inverse* of a classical operation (or a sequence thereof) does not make sense. On the contrary, all quantum operations are inherently *reversible*. Consider an operation g described by the unitary matrix U. Then, its inverse U^{-1} is efficiently calculated as the conjugate-transpose $U^{\dagger}$. Given a sequence of m operations $g_0, \ldots, g_{m-1}$ with associated matrices $U_0, \ldots, U_{m-1}$, the inverse of the corresponding system matrix $U = U_{m-1} \cdots U_0$ is derived by reversing the order of operations and inverting each individual operation, i.e.,

$$U^{-1} = U^\dagger = U_0^\dagger \cdots U_{m-1}^\dagger. \tag{14.1}$$

Now consider two quantum circuits G and G'. The inherent reversibility of quantum computations allows reformulating the equivalence checking problem as the question of whether

$$U \cdot U'^\dagger = e^{i\theta} I, \tag{14.2}$$

with $\theta \in (-\pi, \pi]$ again denoting a physically unobservable global phase. Here, the choice of $U \cdot U'^\dagger$ (in favor of $U' \cdot U^\dagger$, $U^\dagger \cdot U'$, and $U'^\dagger \cdot U$) is arbitrary and interchangeable. Put simply, this formulation states that two quantum circuits are equivalent if the combination of one with the inverse of the other can be reduced to the identity. Since the identity constitutes the best case for many data structures, such as decision diagrams (cf. Sect. 3.1), this offers significant potential.

Unfortunately, creating such a concatenation in a naive fashion, e.g., by computing $U \cdot U'^\dagger$, hardly yields any benefits compared to simply constructing and comparing the respective system matrices. That is, even if the final representation were as compact as possible, the full (and potentially exponential) representations of $U \equiv G$ and $U'^\dagger \equiv G'^{-1}$ would still be generated as intermediate results. Even if the whole sequence of $|G| + |G'|$ operations is considered as one huge quantum circuit whose functionality is to be constructed and compared with the identity, this would still involve the construction of the full representation of one circuit for the first $|G^{(\prime)}|$ operations.

Example 14.1 Consider again the two circuits G and G' from Example 13.1. Concatenating G and G'^{-1} yields a quantum circuit $\tilde{G} = G \cdot G'^{-1}$ with $4 + 16 = 20$ gates. Since both computations are equivalent, constructing the functionality of $\tilde{G}$ would eventually yield the identity. However, during construction, the first four operations would essentially construct the full representation of G. Thus, there is no real benefit to conducting the equivalence check in this way.

Instead, the full potential of this observation is utilized if the associativity of the respective multiplications is fully exploited. More precisely, given two quantum circuits G and G', it holds that

$$
\begin{aligned}
G'^{-1} \cdot G &= (g'^{-1}_{|G'|-1} \cdots g'^{-1}_0) \cdot (g_0 \cdots g_{|G|-1}) \\
&\equiv (U_{|G|-1} \cdots U_0) \cdot (U_0'^\dagger \cdots U_{|G'|-1}'^\dagger) \\
&= U_{|G|-1} \cdots U_0 \cdot I \cdot U_0'^\dagger \cdots U_{|G'|-1}'^\dagger \\
&=: G \to I \leftarrow G'. \tag{14.3}
\end{aligned}
$$

Here, $G \to I \leftarrow G'$ symbolizes that, starting from the identity I, either gates from G can be "applied from the left", or (inverted) gates of G' can be "applied from the right". If the respective gates of G and G' are applied in a fashion that frequently yields the identity, the entire equivalence checking process can be conducted with only rather small (intermediate) representations. This is illustrated by the following example.

Example 14.2 Consider again the circuits G and G' from Example 14.1 and assume that decision diagrams are used as the underlying data structure for the equivalence check. If, starting with a decision diagram representing the identity, gates from both circuits are applied in the sequence

$$\mathbf{I}-g_0-g_0'^\dagger-g_1-g_1'^\dagger-g_2'^\dagger-g_3'^\dagger-g_4'^\dagger-g_2-g_3-g_5'^\dagger-g_6'^\dagger-g_7'^\dagger-g_8'^\dagger-g_9'^\dagger-g_{10}'^\dagger-g_{11}'^\dagger-g_{12}'^\dagger-g_{13}'^\dagger-g_{14}'^\dagger-g_{15}'^\dagger$$

$$(14.4)$$

the number of nodes in the corresponding decision diagrams evolves as

$$3-3-3-4-3-3-4-4-6-6-7-7-7-5-5-4-3-3-3-3-3 \tag{14.5}$$

As can be seen, applying the gates from G and G' in a particular order "from the left" and "from the right", respectively, frequently yields situations where the impact of a gate from circuit G (potentially increasing the size of the decision diagram) is reverted by multiplications with inverted gates from G' (potentially decreasing the size of the decision diagram back to the representation of the identity function).

Moreover, even if the considered circuits G and G' are *not* functionally equivalent (and, hence, identity is not achieved), the observations above still promise improvements compared to creating the complete representations for G and G'. This is because in this case, the result of $G \to I \leftarrow G'$ inherently provides an efficient representation of the circuit's difference that allows one to obtain counterexamples almost "for free" (while state-of-the-art solutions have to explicitly create those using additional inversion and multiplication operations.

Overall, equivalence checking of two quantum circuits can be conducted much more efficiently and compactly than before by exploiting this characteristic. But determining when to apply gates from G and when to apply (inverted) gates from G' is not obvious. Corresponding general-purpose strategies will be presented in Sect. 14.2, while a special-purpose strategy for verifying the results of compilation flows will be presented later in Chap. 15.

14.1.2 Potential Power of Simulation

The second characteristic we are exploiting rests on the observation that simulation is much more powerful for equivalence checking of quantum circuits than for equivalence checking of classical circuits. More precisely, in the classical realm, it is certainly possible to simulate two circuits with random inputs to obtain counterexamples in the event that they are not equivalent. However, this often does not yield the desired result. In fact, due to masking effects and the inevitable information loss introduced by many classical gates, the chance of detecting differences in circuits within a few arbitrary simulations is greatly reduced (e.g., $x \wedge 0$ masks any difference that potentially occurs during the calculation of x). Consequently, sophisticated schemes for constraint-based stimuli generation [4–7], fuzzing [8, 9], etc. are used to verify classical circuits.

This is significantly different in quantum computing. Here, the inherent reversibility of quantum operations dramatically reduces these effects and frequently yields situations where even small differences remain unmasked and affect entire system matrices—showing the power of random simulations for checking the equivalence of quantum circuits. Because of that, it is in general not necessary to compare the *entire* system matrices—in particular when two circuits are *not* equivalent and, hence, their system matrices differ from each other. Then, rather than constructing the overall matrices U and U' for both computations, it is sufficient to simply compare a couple of individual columns. If any of those columns differ, the two circuits have been shown to be *non-equivalent*. This is illustrated by the following example.

Example 14.3 Consider again the circuit G and the erroneous circuit $\tilde{G}'$ from Example 13.1. Upon comparison of the respective system matrices U and $\tilde{U}'$, the non-equivalence of both circuits is clearly seen. Furthermore, since U and $\tilde{U}'$ differ in *all* their columns, this non-equivalence can also be detected by constructing *any* two columns $|u_i\rangle$ and $|\tilde{u}_i'\rangle$, rather than constructing the entire matrices U and $\tilde{U}'$. Specifically, it holds that $\mathcal{F}(|u_i\rangle, |\tilde{u}_i'\rangle) \approx 0.92$ for all i from 0 to 7.

While the construction of U (and accordingly of U') requires expensive matrix-matrix multiplications $U_{m-1} \cdots U_0$, the construction of a single column $|u_i\rangle$ with $i \in \{0, \ldots, 2^n - 1\}$ equates to simulating G with the computational basis state $|i\rangle$ as input, i.e., performing the matrix-vector multiplications

$$|u_i^{(0)}\rangle = U_0|i\rangle, \quad |u_i^{(j)}\rangle = U_j \cdot |u_i^{(j-1)}\rangle \text{ for } j \in \{1, \ldots, |G| - 1\}. \tag{14.6}$$

If the results of those simulations (respectively yielding the ith columns $|u_i\rangle = |u_i^{(|G|-1)}\rangle$ and $|u_i'\rangle = |u_i'^{(|G'|-1)}\rangle$) differ, a counterexample is found and the two circuits have been shown to be *non-equivalent*.

This constitutes an exponentially easier task than constructing the entire system matrices U and U'—although the complexity of simulation still remains exponential with respect to the number of qubits. Regarding the complexity, creating the entire system matrices corresponds to simulating the respective circuit with all 2^n different computational basis states. All this, of course, does not guarantee that any difference is indeed detected simply by simulating a limited number of arbitrary computational basis states $|i\rangle$. But motivated by these observations, a more detailed consideration of this direction was triggered, the results of which are summarized in Sect. 14.3. In combination with the ideas from Sect. 14.1.1, this eventually leads to the proposal of an advanced equivalence checking flow (described in detail in Sect. 14.4) which, first, quickly checks for a possible non-equivalence with some simulation runs. If no counterexample has been obtained by this, the (highly probable) equivalence is proved afterward.

14.2 Exploiting the Power of Reversibility

Following the general ideas outlined in Sect. 14.1.1 potentially allows one to conduct equivalence checking of quantum circuits in a significantly more efficient fashion than before. However, to fully exploit the potential of the idea, a "good" strategy for how to eventually conduct $G \rightarrow I \leftarrow G'$ (i.e., when to apply gates from G and when to apply inverted gates from G') is essential. In this section, we propose several promising strategies and illustrate their application.

14.2.1 Naive Strategy

The first strategy is motivated by the (rather *naive*) assumption that a given circuit G is checked against itself, that is, $G \rightarrow I \leftarrow G$. Then, obviously, the best possible strategy is to alternate between applications of gates from G and their respective inverse—yielding the identity function after each pair of operations. If $G \neq G'$ (and, without loss of generality, assuming that $|G| < |G'|$, that is, that G' has more gates), this strategy alternates between G and G' until all gates of G have been applied. Afterwards, the "left-over" gates from G' are applied. This strategy supposedly works well if G and G' are very similar, but obviously loses its benefits if both circuits significantly differ in their structure (in particular if one circuit has significantly more gates than the other, i.e., if $|G| \ll |G'|$).

Example 14.4 Consider again the two circuits G and G' as discussed before in Example 14.2. Applying the naive strategy leads to an order of gate applications, as shown at the top of Fig. 14.1 with the corresponding node count listed below the respective identifier. During this process, the size of the (intermediate) decision diagrams never exceeds 7 nodes, while the average node count is 4.43. This is significantly less than required by the (established)

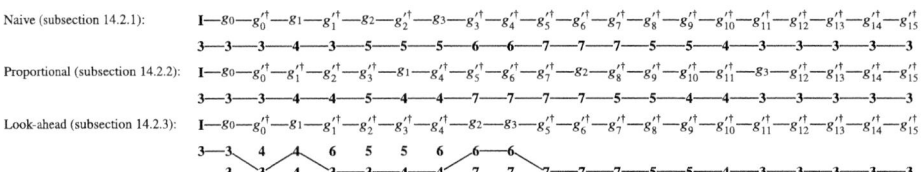

Fig. 14.1 Application of the proposed strategies and corresponding node counts (for the circuits G and G' from Example 13.1)

approach that constructs and compares individual decision diagrams, which has a maximal node count of 9 and an average one of 5.95.

14.2.2 Proportional Strategy

Applying the naive strategy to circuits that, structurally, are significantly different obviously leads to an imbalance since a huge portion of "left-over" gates are applied—possibly neglecting the effect of staying close to the identity function. To avoid this, the *proportional* strategy aims for a more balanced approach. To this end, the ratio with respect to the number of gates for both circuits is first determined. Subsequently, the gates from G and G' are proportionally applied according to this ratio.

Example 14.5 Consider again the two circuits G and G' as discussed before in Example 14.2. The ratio between their gate counts is $4 : 16 = 1 : 4$. Hence, applying the proportional strategy leads to an order of gate applications, as shown in the second row of Fig. 14.1, with the corresponding node count listed below the respective identifier. During this process, the size of the (intermediate) decision diagrams never exceeds 7 nodes, while the average node count is 4.33. For the problem at hand, this strategy constitutes near-optimal performance.

14.2.3 Look-Ahead Strategy

Despite strategies considering structural elements of the given circuits to decide the actions to be performed, also schemes based on the actual size of the (intermediate) representations may provide a good indication of how to proceed. Recall that the general aim is to stay as close to the identity function as possible (e.g., leading to the smallest possible decision diagram). Hence, the decision to apply a gate from G or G' can be based on which case actually leads to a smaller representation. This is conducted by the *look-ahead* scheme. While this potentially doubles the number of multiplications to be performed (since both alternatives have to be checked out for each gate), it may lead to smaller representations and, by this, to a more efficient equivalence checking routine.

Example 14.6 Consider again the two circuits G and G' as discussed before in Example 14.2. Applying the look-ahead strategy leads to an order of gate applications, as shown in the third row of Fig. 14.1. As long as there are still gates left to multiply in *both* circuits, the bottom of Fig. 14.1 indicates the node count of both alternatives (G at the top, G' at the bottom). The bold line indicates which path was chosen. The resulting sequence of operations is exactly the one used in Example 14.2—resulting in a maximum of 7 nodes, while the average node count is 4.24.

Even for the small example showcased throughout this section, all proposed strategies perform significantly better in terms of maximum as well as average intermediate representation size when compared to the established approach of constructing and comparing individual decision diagrams.

14.3 Exploiting the Power of Simulation

In Sect. 14.1.2, we illustrated the potential power of simulation for equivalence checking and argued that, in case two quantum circuits G and G' operating on n qubits are *not* equivalent, even a few simulation runs will likely yield a counterexample. This idea triggered a more detailed consideration in which we elaborated how significantly the matrices U and U' differ from each other in case of non-equivalence and whether this would make an incomplete coverage of the functionality feasible. The results obtained by this consideration are summarized in this section.

To this end, we first introduce the notion of the *difference* of two unitary matrices. Given two unitary matrices U and U', their *difference* D is defined as the unitary matrix $D = U^\dagger U'$ and it holds that $U \cdot D = U'$. Similarly to picking $G'^{-1} \cdot G$ in Sect. 14.1.1, the particular arrangement $D = U^\dagger U'$ is just one of the four possibilities to introduce the notion of the *difference* of two unitaries. In case both matrices are identical (i.e., the circuits are equivalent), it directly follows that $D = I$. One characteristic of the identity function I resulting in this case is that all diagonal entries are equal to one, i.e., $\langle i | U^\dagger U' | i \rangle = 1$ for $i \in \{0, \ldots, 2^n - 1\}$, where $|i\rangle$ denotes the ith computational basis state. More generally, in case of a potential relative/global phase difference between G and G', all diagonal elements have modulus one, i.e., $|\langle i | U^\dagger U' | i \rangle|^2 = 1$. This expression can further be rewritten to

$$1 = |\langle i|U^\dagger U'|i\rangle|^2 = |(U|i\rangle)^\dagger (U'|i\rangle)|^2 = ||u_i\rangle^\dagger |u_i'\rangle|^2 = |\langle u_i | u_i' \rangle|^2, \qquad (14.7)$$

where $|u_i\rangle$ and $|u_i'\rangle$ denote the ith column of U and U', respectively. This essentially resembles the simulation of both circuits with the initial state $|i\rangle$ and, later, calculating the fidelity $\mathcal{F}$ between the resulting states $|u_i\rangle$ and $|u_i'\rangle$. Hence, if only one simulation yields $\mathcal{F}_i := \mathcal{F}(|u_i\rangle, |u_i'\rangle) \not\approx 1$, then $|i\rangle$ proves the non-equivalence of G and G'.

Now, the question is how many computational basis states $|i\rangle$ yield $\mathcal{F}_i \not\approx 1$ for a given difference matrix D, i.e., how likely it is for an arbitrary simulation to detect possible differences. Since the difference D of both matrices is unitary itself, it can also be interpreted as a quantum circuit G_D. For the purpose of this theoretical consideration, we assume that each gate of G_D represents either a single-qubit or a (multi-) controlled operation[1].

---

[1] This does not limit the applicability of the following findings, since arbitrary single-qubit operations combined with CNOT form a universal gate-set [10].

Example 14.7 Assume that G_D consists only of one (non-trivial) single-qubit operation defined by the matrix U_s applied to the first of n qubits. Then, the system matrix D is given by

$$D = I_{2^{n-1}} \otimes U_s = \begin{bmatrix} U_s & & \\ & \ddots & \\ & & U_s \end{bmatrix}. \tag{14.8}$$

The process of going from U to U', i.e., calculating $U \cdot D$, impacts *all* columns of U. Thus, an error is detected by a *single* simulation with *any* computational basis state.

Among all quantum operations, single-qubit operations possess a system matrix that is least similar to the identity matrix due to the tensor product structure of their corresponding system matrix.

Example 14.8 In contrast to Example 14.7, assume that G_D consists of only one operation U_s targeted at the first qubit and controlled by the remaining $n - 1$ qubits. Then, the corresponding system matrix is given by

$$D = \begin{bmatrix} I_2 & & & \\ & \ddots & & \\ & & I_2 & \\ & & & U_s \end{bmatrix}. \tag{14.9}$$

In this case, applying D to U only affects the last two columns of U. As a consequence, a maximum of two columns (out of 2^n) may serve as counterexamples—the worst case scenario.

These basic examples cover the extreme cases when it comes to the difference of two unitary matrices. In case G_D does not exhibit such a simple structure, the analysis is more involved. Generally, quantum operations with $c \in \{0, \dots, n - 1\}$ control qubits will exhibit a difference in at most 2^{n-c} columns. Furthermore, given two operations showing a certain number of differences, the matrix product of these operations in most cases (except when cancellations occur) differs in as many columns as the maximum of both operands.

Example 14.9 Consider a two-qubit system and two circuits G and G' exhibiting a difference circuit G_D consisting of two gates—the first of which is a Hadamard operation $H(q_1)$ on the second qubit, while the second is a *CNOT* operation $CNOT(q_1, q_0)$ with target on the first and control on the second qubit. As discussed in Examples 14.7 and 14.8, the application of the single-qubit matrix U_H affects all columns, while the application of the controlled two-qubit operation matrix U_{CNOT} only affects two of the four columns. Their combination, however, still affects *all* columns of the result.

The gate-set provided by (current) quantum computers typically includes only (certain) single-qubit gates and a specific two-qubit gate, such as the *CNOT* gate. Thus, multi-controlled quantum operations usually only arise at the most abstract algorithmic description of a quantum circuit and are then *decomposed* into elementary operations from the device's gate-set before the circuit is mapped to the target architecture. As a consequence, errors occurring during the design flow will typically consist of (1) single-qubit errors, e.g., offsets in the rotation angle, or (2) errors related to the application of *CNOT* or *SWAP* gates. In both cases, non-equivalence can be efficiently concluded by a limited number of simulations with arbitrary computational basis states.

Of course, this cannot be guaranteed. Otherwise, the problem would not be QMA-complete. However, following the above discussion and considering that comparing single columns (i.e., a partial consideration of the functionality) is substantially cheaper than a full functional coverage, simulation of arbitrary computational basis states is a promising option. Moreover, if a counterexample was not obtained after a few simulations, this yields a highly probable estimate of the circuit's equivalence—in contrast to the classical realm, where this generally does not allow for any conclusion.

14.4   Resulting Advanced Equivalence Checking Flow

The general ideas proposed in Sect. 14.1 and elaborated in Sects. 14.2 and 14.3 complement each other in many different ways. Trying to keep $G \rightarrow I \leftarrow G'$ close to the identity (see Sect. 14.1.1) proves very efficient in case two circuits are indeed equivalent—provided a "good" strategy can be employed. On the other hand, conducting simulations with randomly chosen computational basis states $|i\rangle$ (see Sect. 14.1.2) allows one to quickly detect non-equivalence even in cases where both circuits differ only slightly. These considerations eventually lead to an advanced equivalence checking flow as shown in Fig. 14.2. Here, instead of constructing and comparing the complete matrix representations of both circuits, we propose first to perform a limited number of $r \ll 2^n$ simulation runs with randomly chosen computational basis states. Should one of those simulations yield different outputs in both circuits (i.e., a fidelity $\mathcal{F}_i \not\approx 1$), the non-equivalence of the circuits under consideration has been shown. If

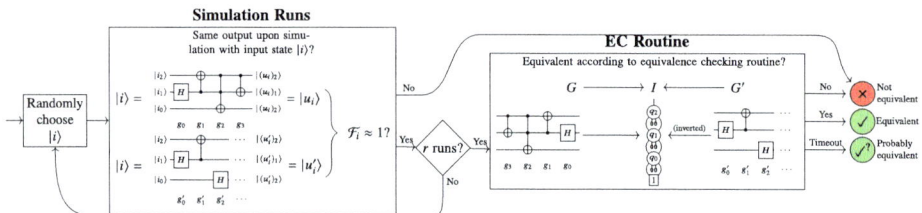

Fig. 14.2 Proposed equivalence checking flow

this is not the case, the equivalence checking routine $G \rightarrow I \leftarrow G'$ is used to complete the task. Moreover, if simulation has not revealed any differences, the likelihood of the circuits being non-equivalent is significantly reduced, as shown in the discussions from Sect. 14.3.

Overall, this eventually leads to three possible outcomes:

- *Not equivalent*, i.e., if any simulation run yields $\mathcal{F}_i \not\approx 1$ or if the equivalence checking routine is employed after the simulation runs and yields this result.[2] In many cases, this can be conducted very efficiently, while existing equivalence checking routines require substantial runtime or even time out frequently.
- *Equivalent*, i.e., if, after r simulation runs, the equivalence checking routine is employed and yields that result. If this is the case, the simulation runs conducted before (which did not lead to a conclusive result) only constitute a negligible runtime overhead. As discussed earlier, this is because constructing the whole functionality is complexity-wise equivalent to simulating *all* 2^n computational basis states, and we have chosen $r \ll 2^n$.
- *Timeout*, i.e., if the simulation runs did not lead to a counterexample and the equivalence checking routine was unable to complete the task within a given resource limit. If this is the case, we at least get an indication that both circuits might be equivalent (since the conducted simulations did not provide a counterexample, which, according to the discussions from Sect. 14.3, is rather rare). Even if this does not provide a guarantee of non-equivalence, this is a stronger result than provided by the state of the art thus far, which does not allow for any kind of conclusion in this case.

The resulting equivalence checking flow with the combination of using the power of simulation together with the dedicated $G \rightarrow I \leftarrow G'$ scheme dramatically improves the state of the art. In contrast to constructing and comparing the whole functionality of two circuits, simulation can typically be conducted significantly faster while already providing valuable insights, i.e., a counterexample or an indication that both circuits might be equivalent. If the results of the simulation remain inconclusive, that is, no counterexample was found, the subsequent functional comparison can be improved by using the $G \rightarrow I \leftarrow G'$ scheme. Even if the latter step (the actual proof) cannot be completed due to time or resource limits, a meaningful result has already been obtained after conducting the simulations—namely that the circuits are presumably equivalent—while previously proposed approaches provide no indication on the equivalence at all.

Naturally, there are cases where even single simulation runs prove too resource-consuming to conduct. Especially in cases where simulation fails due to the immense complexity of the involved state vector, the $G \rightarrow I \leftarrow G'$ scheme still provides a complementary alternative. If a strategy can be employed which manages to keep the intermediate representations close to the identity, the whole procedure can again be conducted with a rather low memory footprint. The design of corresponding strategies could benefit from specific knowledge about

---

[2] As a consequence of the discussions from Sect. 14.3 it is far more likely that this result already occurs during simulation.

the origin of both circuits, e.g., if G' is the result of compiling G to a certain target device. Such a strategy is proposed in Chap. 15.

In addition to that, our discussions in Sect. 14.3 show that choosing computational basis states uniformly at random as input for the simulation part of the proposed equivalence checking flow promises high success rates in the quantum realm. This is in stark contrast to the (simulative) verification of classical circuits, where sophisticated techniques such as constrained-based stimuli generation [4–7], fuzzing [8, 9], etc. must be employed for such techniques to work at all. The first results towards the study of quantum stimuli generation schemes can be found later in Chap. 16.

Furthermore, previous work on equivalence checking of quantum circuits often used several *optimizations* such as template replacement [11, 12]. Potential candidates include optimizations frequently used during compilation [12–16]. Those optimization techniques may of course also be applied in the methodology proposed here—as long as the optimization itself is valid (which could be verified beforehand using the proposed methodology). In general, lowering the gate count of the respective circuits (or their concatenation in the case of the $G \to I \leftarrow G'$ scheme) can be expected to improve the performance of the proposed equivalence checking flow (since less operations have to be applied).

Finally, it is not necessary to perform both complementary steps (simulation and the $G \to I \leftarrow G'$ scheme) sequentially, as proposed in Fig. 14.2. In fact, the r simulations and the $G \to I \leftarrow G'$ scheme can also be started in parallel (provided sufficient memory and processor resources). If then any of the simulations yields a counterexample, the remaining calculations can be aborted. Vice versa, if the equivalence checking routine returns "equivalent", any ongoing simulation runs can be stopped.

14.5   Summary of Results

The proposed methodology has been implemented and serves as the basis for the QCEC quantum circuit equivalence checking tool [17], which is part of the *Munich Quantum Toolkit* (MQT, [18]). QCEC is open source and publicly available at https://github.com/munich-quantum-toolkit/qcec.

Its core is written in C++ and is based on the MQT decision diagram package, which was co-developed as part of this book. To make the tool as user-friendly as possible, it includes easy-to-use Python bindings, is offered as pre-built Python wheels for all major platforms, and interfaces easily with IBM's Qiskit.

All of the strategies proposed in this chapter have been thoroughly evaluated in [1–3], which led to the following results:

- In the majority of cases, the dedicated strategies proposed in this chapter reduce the equivalence checking time down to a half or a third compared to the established routine that constructs and compares the respective decision diagrams. The additionally conducted simulations only lead to a negligible runtime overhead.
- The proposed methodology can determine the non-equivalence for all considered benchmarks in just a matter of seconds or even less, whereas the existing technique requires substantial runtime or does not finish within the given time limit at all. Moreover, in the vast majority of cases, simulation alone is capable of showing the non-equivalence.
- A third evaluation on the success probability of the simulation part clearly shows the power of simulation to either quickly show the non-equivalence of two circuits or, at least, provide an indication of equivalence. In particular, for cases where no decisive answer could be obtained by any equivalence checking routine due to timeouts, this is better than no result at all.

Overall, these experimental results demonstrate that the resulting methodology allows for much faster equivalence checking than ever before. In many cases, a single simulation run is sufficient.

Implementation, Usage, Documentation, and Results

 The proposed methodology is available as part of the open-source MQT QCEC tool [17] at https:// github . com/ munich-quantum-toolkit/ qcec. which can be installed using pip install mqt . qcec.

 Using QCEC to verify the equivalence of two Qiskit quantum circuits is as easy as:

```
from mqt import qcec
from qiskit import QuantumCircuit

qc1 = QuantumCircuit(2)
qc1.cx(0, 1)

qc2 = QuantumCircuit(2)
qc2.h(0)
qc2.h(1)
qc2.cx(1, 0)
qc2.h(1)
qc2.h(0)

qcec.verify(qc1, qc2)
```

 Documentation on all available configuration options is available at https : //mqt . readthedocs . io/projects/qcec

 Details on the experimental setup, evaluations, and results can be found in [1]–[3].

References

1. L. Burgholzer, R. Wille, Advanced equivalence checking for quantum circuits. IEEE Trans. CAD Integr. Circuits Syst. (2021). https://doi.org/10.1109/TCAD.2020.3032630
2. L. Burgholzer, R. Wille, Improved DD-based equivalence checking of quantum circuits, in *Asia and South Pacific Design Automation Conference* (2020)
3. L. Burgholzer, R. Wille, The power of simulation for equivalence checking in quantum computing, in *Design Automation Conference* (2020)
4. J. Yuan, C. Pixley, A. Aziz, *Constraint-Based Verification* (Springer, Berlin 2006)
5. J. Bergeron, *Writing Testbenches Using System Verilog* (Springer, Berlin, 2006)
6. N. Kitchen, A. Kuehlmann, Stimulus generation for constrained random simulation, in *Int'l Conference on CAD* (2007), pp. 258–265
7. R. Wille, D. GroSSe, F. Haedicke, R. Drechsler, SMT-based stimuli generation in the SystemC Verification library, in *Forum on Specification and Design Languages* (2009)
8. K. Laeufer, J. Koenig, D. Kim, J. Bachrach, K. Sen, RFUZZ: coverage-directed fuzz testing of RTL on FPGAs, in *Int'l Conference on CAD* (2018)
9. H.M. Le, D. GroSSe, N. Bruns, R. Drechsler, Detection of hardware Trojans in SystemC HLS designs via coverage-guided fuzzing, in *Design, Automation and Test in Europe* (2019)
10. M.A. Nielsen, I.L. Chuang, *Quantum Computation and Quantum Information* (Cambridge University Press, 2010)
11. S. Yamashita, I.L. Markov, Fast equivalence-checking for quantum circuits, in *Int'l Symposium on Nanoscale Architectures* (2010). https://doi.org/10.1109/NANOARCH.2010.5510932
12. R. Duncan, A. Kissinger, S. Perdrix, J. van de Wetering, Graph-theoretic simplification of quantum circuits with the ZX-calculus (2019). arXiv:1902.03178, preprint
13. A.K. Prasad, V.V. Shende, I.L. Markov, J.P. Hayes, K.N. Patel, Data structures and algorithms for simplifying reversible circuits. J. Emerg. Technol. Comput. Syst. 2(4), 277–293 (2006)
14. K. Iwama, Y. Kambayashi, S. Yamashita, Transformation rules for designing CNOT-based quantum circuits, in *Design Automation Conference* (2002), pp. 419–424
15. D. Maslov, G. Dueck, D. Miller, C. Negrevergne, Quantum circuit simplification and level compaction. IEEE Trans. CAD Integr. Circuits Syst. 27(3), 436–444 (2008). ISSN: 0278-0070, 1937-4151. https://doi.org/10.1109/TCAD.2007.911334
16. G. Vidal, C.M. Dawson, Universal quantum circuit for two-qubit transformations with three controlled-NOT gates. Phys. Rev. A 69(1), 010 301 (2004). ISSN: 1050-2947, 1094-1622. https://doi.org/10.1103/PhysRevA.69.010301
17. L. Burgholzer, R. Wille, QCEC: a JKQ tool for quantum circuit equivalence checking. Softw. Impacts (2021)
18. R. Wille, L. Berent, T. Forster, et al., The MQT handbook: a summary of design automation tools and software for quantum computing, in *IEEE International Conference on Quantum Software (QSW)* (2024). https://doi.org/10.1109/QSW62656.2024.00013. arXiv:2405.17543, A live version of this document is available at https://mqt.readthedocs.io

In this chapter (based on [1]), we propose a scheme for *quantum circuit equivalence checking* that is optimized to verify the results of the compilation flows of quantum circuits. To this end, we create a dedicated equivalence checking strategy by combining quantum-specific characteristics based on the methodology proposed in Chap. 14 and knowledge of the compilation flow. By utilizing information about the compilation flow, we demonstrate that the respective intermediate representations can be kept close to the identity in an almost perfect fashion. This frequently reduces exponential complexity to linear or near-linear complexity, thereby significantly reducing the runtime of the verification process.

The remainder of this chapter is structured as follows: Sect. 15.1 elaborates on the problem of verifying the results of compilation flows. Then, Sect. 15.2 illustrates how different aspects of the compilation flow can be exploited and, consequently, orchestrated to form a dedicated verification strategy. Section 15.3 concludes with a discussion of the resulting implementation and a summary of the experimental results.

15.1 Verifying Results of the Compilation Flow

The quantum circuit compilation flow (as reviewed in Chap. 9) converts a high-level quantum circuit into a representation that adheres to all restrictions imposed by the target device. Verifying the results of this procedure involves ensuring that the resulting circuit preserves the functionality that was originally intended. Therefore, the verification of an original circuit G and a compiled circuit G' is reduced to determining whether both circuits are equivalent. To do this, the $G \rightarrow I \leftarrow G'$ methodology proposed in Chap. 14 can be used directly. Recall that the intention of the concept is to alternate between gate applications from both circuits so that the intermediate representations stay as close to the identity as possible during the

L. Burgholzer and R. Wille, *Design Automation Tools and Software for Quantum Computing*, https://doi.org/10.1007/978-3-032-06770-8_15

computation. Since the identity is the optimal case for most representations of quantum functionality (e.g., linear in the number of nodes with respect to the number of qubits for decision diagrams), this constitutes an efficient equivalence checking method.

Example 15.1 Assume, without loss of generality, that $|G| \leq |G'|$, i.e., G' has at least as many gates as G. Further assume an *oracle* $\omega\colon G \to (G')^*$ exists that, given a gate $g_i \in G$, returns a consecutive sequence of gates $g'_k \ldots g'_l \in G'$ such that $g_i \equiv g'_k \ldots g'_l$. Then, subsequently applying one gate $g \in G$ and $|\omega(g)|$ inverted gates from G' constitutes a "perfect" strategy for conducting the equivalence check—yielding the identity after each pair of applications. Thus, only matrices that remain close to the identity occur. Since these can usually be represented very efficiently using, e.g., decision diagrams, the process of equivalence checking is substantially improved.

However, the greatest obstacle is determining when to apply operations of G ("from the left") and when to apply operations of G' ("from the right") in order to obtain the "perfect" oracle $\omega(\cdot)$. Several strategies for this purpose were proposed in Chap. 14, and significant speed-ups in checking the equivalence of two quantum circuits were achieved with these strategies [2]. However, these strategies remain relatively simple and problem agnostic (e.g., they use a one-to-one or size-proportional application of gates from G and G'). As such, they hardly resemble a "perfect" strategy that can keep the computation of $G \to I \leftarrow G'$ close to the identity in case of verifying compilation flow results. However, the compilation flow (as reviewed in Chap. 9) provides detailed insights into how a circuit G is ultimately compiled into a circuit G', thereby providing the knowledge necessary to work towards a "perfect" oracle $\omega(\cdot)$. In the following, a verification scheme is proposed that uses the idea of applying $G \to I \leftarrow G'$ and, at the same time, uses knowledge of actual compilation flows to derive a much better oracle $\omega(\cdot)$.

15.2   Proposed Verification Scheme

In this section, we show how knowledge about each individual step in the compilation flow—specifically, synthesis, mapping, and optimization—can contribute to an oracle $\omega(\cdot)$ that keeps the intermediate representations as close to the identity as possible during the computation of $G \to I \leftarrow G'$. As confirmed by the evaluations (summarized in Sect. 15.3), this allows for significant speed-ups and makes large-scale verification of compilation results practical.

15.2.1 Using Knowledge About the Synthesis Step

The goal of synthesis routines is to transform the gates of the original quantum circuit to the native gate set that the target device supports. To this end, two issues become relevant for determining the "perfect" $G \rightarrow I \leftarrow G'$ strategy:

1. each gate $g \in G$ is compiled to a sequence of gates $g'_k \ldots g'_l \in G'$ and
2. the circuits G and G' may operate on different numbers of qubits due to ancillary qubits required for the synthesis.

For the first issue, it can be exploited that the actual decomposition scheme, i.e., into how many elementary gates each of the original circuit's gates is decomposed, is known a priori. Thus, an *oracle* $w(\cdot)$ which, given a gate $g \in G$, returns the corresponding sequence of gates $g'_k \ldots g'_l \in G'$, is explicitly known. Assuming that G' resulted from the synthesis of a given quantum circuit G, applying *one* gate from G and $|w(g)|$ inverted gates from G' constitutes an optimal strategy for conducting $G \rightarrow I \leftarrow G'$—yielding the identity after each step.

Example 15.2 Consider the following circuit G, representing an instance of Grover's algorithm:

$$(15.1)$$

Executing this circuit on any existing quantum computing platform at least requires decomposing the three-qubit Toffoli gate into elementary gates. A well-known decomposition into single- and two-qubit gates is given by the following:

$$(15.2)$$

Thus, assuming that the targeted device supports arbitrary single-qubit gates and CNOT gates, $|w(g)| = 1$ holds for all $g \in G$ except for the Toffoli gate, where $|w(g)| = 15$ holds.

In the situation where both circuits operate on a different number of qubits, the typical formulation of the equivalence verification problem presented in Sect. 13.2 cannot be used directly. This is because quantum circuits with different numbers of qubits cannot represent the same unitary. Actually, their unitaries are not even the same size. Sadly, it is not sufficient to match the qubit count of G' by adding idle qubits to the original circuit. Since ancillary qubits are always presumed to be initialized in a specific state (usually $|0\rangle$), the overall unitary representation U' retains some degree of freedom. To accommodate for this degree of freedom, the problem formulation must be slightly adjusted by fixing the ancillary qubits in the circuits to their constant state. An example illustrates the idea.

Example 15.3 Consider a unitary $2^n \times 2^n$ matrix U and assume that, without loss of generality, the last qubit q_{n-1} acts as an auxiliary qubit initialized to $|0\rangle$. In general, the action of U depending on the state of q_{n-1} is described by four $2^{n-1} \times 2^{n-1}$ sub-matrices U_{ij}, namely:

$$U: \; \begin{array}{c} \\ \end{array} \begin{array}{cc} & \text{from} \\ q_{n-1} & |0\rangle \quad |1\rangle \\ \begin{array}{c} |0\rangle \\ |1\rangle \end{array} \left[\begin{array}{c|c} U_{00} & U_{10} \\ \hline U_{01} & U_{11} \end{array} \right] \end{array} \tag{15.3}$$

Since the ancillary is initialized to $|0\rangle$, the sub-matrices corresponding to the transformation from $|1\rangle$ can be ignored—resulting in the following *modified matrix*:

$$\tilde{U}: \; \begin{array}{cc} & \text{from} \\ q_{n-1} & |0\rangle \quad |1\rangle \\ \begin{array}{c} |0\rangle \\ |1\rangle \end{array} \left[\begin{array}{c|c} U_{00} & 0 \\ \hline U_{01} & 0 \end{array} \right] \end{array} \tag{15.4}$$

If, additionally, the ancillary is guaranteed to return to its initial state at the end of the computation (in this case $|0\rangle$), only the unitary matrix U_{00} must be considered in the equivalence check.

Overall, these techniques allow one to determine a perfect oracle $\omega(\cdot)$ for the synthesis step and handle scenarios where both circuits act on a different number of qubits.

15.2.2 Using Knowledge About the Mapping Step

Mapping to the targeted architecture establishes a connection between the circuit's logical and the device's physical qubits.[1] Consequently, while the description of G is expressed in terms of logical qubits $q_0, \ldots, q_{n-1}$, the circuit G' operates on (a subset of) the device's physical qubits $Q_0, \ldots, Q_{N-1}$. If a non-trivial initial mapping (i.e., anything but $q_i \mapsto Q_i$) is employed, this leads to the situation that gates from G', although functionally equivalent, are applied to different qubits than the gates of G. Therefore, concluding the equivalence of both circuits is not possible simply by using the oracle function $\omega(\cdot)$. Instead, a *qubit map* $m(\cdot)$ is used, which stores the mapping between the physical qubits of the circuit G' and the logical qubits of the original circuit G, i.e., $m(Q_i) = q_j$ if physical qubit Q_i is initially assigned logical qubit q_j. Whenever a gate from G' is applied to a certain physical qubit

---

[1] The terms *logical* and *physical* qubit are used here to denote levels of abstraction and should not be confused with notions from, e.g., quantum error correction. For a quantum circuit operating on n qubits that shall be executed on a device that provides $N \geq n$ qubits, we refer to the n qubits of the circuit as the *logical* qubits and to the N qubits of the device as *physical* qubits. See also Chap. 9.

Q_i, this is translated to the corresponding logical qubit $m(Q_i) = q_j$—again allowing one to stay close to the identity.

Example 15.4 Consider the circuit G from Example 15.2 and the following mapped version $\tilde{G}$:

$$(15.5)$$

While the X gate at the beginning of G is applied to the logical qubit q_2, it is applied to the physical qubit Q_1 in the circuit $\tilde{G}$. In order to fix this mismatch, the qubit map $m(\cdot)$ that maps $Q_0 \mapsto q_0$, $Q_1 \mapsto q_2$, and $Q_2 \mapsto q_1$ is employed. Consequently, the X gate of $\tilde{G}$ is applied to $m(Q_1) = q_2$ which now matches the original gate from G perfectly.

However, the logical-to-physical qubit mapping of a compiled circuit in general changes dynamically throughout the circuit to satisfy all constraints imposed by the device's coupling map. As a consequence, the potential to use the (static) qubit map $m(\cdot)$ in combination with the oracle function $w(\cdot)$ to stay close to the identity is significantly diminished. That is, because the dynamically changed mapping again results in a scenario where the gates from G' are applied to different qubits than in the circuit G. Therefore, a perfect verification strategy must track the changes in the logical-to-physical qubit mapping caused by *SWAP* operations and, accordingly, needs to *update the qubit map $m(\cdot)$* throughout the verification procedure.

Example 15.5 Consider again the scenario of Example 15.4. If the $G \rightarrow I \leftarrow G'$ scheme is carried out using the qubit map $m(\cdot)$ defined there, the result would not represent the identity. That is, because the logical-to-physical qubit mapping is changed in the middle of $\tilde{G}$ by a *SWAP* operation applied to Q_0 and Q_1. Thus, at that specific point, the qubit map $m(\cdot)$ has to be updated accordingly, i.e., it then has to map $Q_0 \mapsto q_2$, $Q_1 \mapsto q_0$, and $Q_2 \mapsto q_1$. Through this dynamic change, the computation of $G \rightarrow I \leftarrow G'$ remains close to the identity and, eventually, proves the equivalence of both circuits.

In general, the introduction of a qubit map $m(\cdot)$ allows one to verify the results of the mapping step. By dynamically updating the map throughout the verification process, the intermediate representations can be kept as close to the identity as possible.

15.2.3 Using Knowledge About the Optimization Step

If no optimizations were applied to the circuit resulting from the synthesis and mapping step (as, e.g., at the Qiskit optimization level $O0$), the strategies proposed above allow one to conduct $G \rightarrow I \leftarrow G'$ in a perfect fashion—yielding the identity after each pair of applications. However, optimizations, which are commonly conducted during compilation to make the circuits more efficient, further alter the circuit, and, as a consequence, make it harder to verify. In the following, we cover how to anticipate the effects of the two most common optimizations:

1. *Single-qubit gate fusion*, where consecutive single-qubit gates placed on a qubit are combined and resynthesized into a shorter sequence of gates, e.g.:

$$\boxed{H}\boxed{X}\boxed{H} \quad \rightarrow \quad \boxed{Z} \tag{15.6}$$

2. *Adjacent gate cancellation*: where redundant sequences of gates are eliminated based on some gate cancellation rules, e.g.:

$$\rightarrow \tag{15.7}$$

Example 15.6 Consider again the circuit $\tilde{G}$ from Example 15.4. There, the blue box indicates the gates of $\tilde{G}$ realizing the Toffoli gate of the original circuit G from Example 15.2. The middle qubit thereby contains a T gate, which is directly followed by an H gate. By performing single-qubit gate fusion, these can be merged into a single $U_2(0, \frac{5\pi}{4})$ gate. Thus, $|w(g)| = 15$ no longer holds but has to be modified to $|w(g)| = 14$ instead in the case of the Toffoli gate.

In addition to anticipating fusions within individual gate realizations through adaptations of the oracle function $w(\cdot)$, a *preprocessing* pass is conducted that fuses consecutive single-qubit gates where they are present in the original circuit G (e.g., fusing the *H-X-H* cascade at the end of the circuit G from Example 15.2 to a single Z gate). However, reductions across multiple gates that were decomposed during synthesis cannot be accounted for in this manner. Thus, the previously constructed *perfect* oracle function $w(\cdot)$ becomes *approximate*.

Example 15.7 Consider $\tilde{G}$ from Example 15.4 and the following optimized variant G':

$$\tag{15.8}$$

Then, the cancellation of the two consecutive H gates in the beginning of $\tilde{G}$ cannot be anticipated through a straightforward adaptation of $w(\cdot)$. However, as also confirmed by experimental evaluations (summarized in Sect. 15.3), $w(\cdot)$ remains a suitable approximation for staying close to the identity.

The second optimization employed per default—adjacent gate cancellation—introduces a peculiar issue for the verification strategy as shown in the following example.

Example 15.8 Consider a *SWAP* operation directly followed by a *CNOT* operation. Since the *SWAP* operation itself is realized by three consecutive *CNOT*s, this sequence of operations may be simplified by cancelling two of them, i.e.:

$$
\begin{array}{c}
q_1 \\
q_0
\end{array}
\equiv
\begin{array}{c}
q_1 \\
q_0
\end{array}
\equiv
\begin{array}{c}
q_1 \\
q_0
\end{array}
\quad ?
\quad ?
\tag{15.9}
$$

While the qubit map $m(\cdot)$ can be easily adapted in the first two cases, the optimized circuit shows no sign of an applied *SWAP* and, furthermore, introduces an additional *CNOT* gate to the compiled circuit which previously did not exist in G. This makes it difficult for the proposed strategies to still identify the *SWAP* and, hence, update the qubit map $m(\cdot)$ as described in the previous section.

As a solution, any occurrence of two consecutive *CNOT* operations in G' as shown on the right-hand side of the above equation that is not followed by a third matching *CNOT* is *substituted* by the sequence shown on the left-hand side of the above equation. Overall, this again allows accurate tracking of the qubit mapping $m(\cdot)$ and the equivalence check to be carried out optimally.

15.2.4 Resulting Verification Scheme

All the above considerations finally result in a dedicated verification scheme that is tailored to verify the results of the compilation flows. First, a preprocessing step fuses subsequent single-qubit gates in the circuit G and substitutes *SWAP* gates (and possibly a *CNOT* gate) where applicable in the circuit G'. Afterward, if necessary, the circuit G is augmented with idle ancillary qubits. Then, the general $G \rightarrow I \leftarrow G'$ scheme is employed—using the oracle function $\omega(\cdot)$ to determine which gates from G' are to be applied for each application of a gate from G.

The actual application of gates from G' occurs with respect to the qubit map $m(\cdot)$ which establishes the connection between the (logical) qubits of the circuit G' and the (physical) qubits of G''. During these steps, this qubit map is dynamically updated to account for the insertion of *SWAP* operations during the mapping. After applying all gates of both circuits, the result of this scheme is modified as illustrated in Example 15.3. Eventually, the two circuits are shown to be equivalent if the modified result resembles the identity for the non-ancillary qubits.

As confirmed by the experimental evaluations, which are summarized next, this scheme allows one to efficiently verify even large instances consisting of tens of thousands of gates within seconds. Additionally, in contrast to formally verifying the individual compilation steps, this approach remains generic enough to work well out of the box, even when optimizations are employed that have not been directly accounted for, e.g., commutation rules.

15.3   Summary of Results

The proposed verification scheme has been integrated into the open-source MQT QCEC tool [3] introduced in Chap. 14. The tool provides native support for verifying the results of IBM Qiskit's quantum circuit compilation flow with dedicated oracle functions $\omega(\cdot)$ for Qiskit's various optimization levels. Extensive experimental evaluations have been performed on the performance of the resulting verification scheme in [1]. This led to the following main results:

- A first series of evaluations considering Qiskit's default optimization level O1 has shown that additionally exploiting explicit knowledge about the compilation flow as proposed in this chapter allows for the verification of *all* considered instances within (fractions of) seconds, whereas existing equivalence checking methods and even the methodology proposed in Chap. 14 frequently time out or require substantial runtime.
- In a second series of evaluations, the proposed method was shown to work almost as well in situations where it was not specifically tailored for. Specifically, the verification of circuits compiled with optimization level O2 was considered, where several more advanced optimization techniques, such as gate commutation rules, are used but not directly accounted for in the proposed scheme.

Overall, the evaluations confirmed that the proposed strategy consistently allows for verifying instances with more than ten thousand gates within seconds—even if optimizations are employed which are not directly accounted for. Compared to the state of the art, which often requires substantial runtimes or even timeouts in these tasks, this is a drastic improvement.

Implementation, Usage, Documentation, and Results

 The proposed verification scheme is available as part of the open-source MQT QCEC tool [3] at https://github.com/munich-quantum-toolkit/qcec, which can be installed using pip install mqt.qcec.

 After defining the circuit G from Example 15.2 in Qiskit and compiling it to an architecture, verifying that the circuit has been compiled correctly is as easy as:

```python
from qiskit import QuantumCircuit, transpile
from qiskit.providers.fake-provider import Generic Backend V2
from mqt import qcec

circ = QuantumCircuit(3)
circ.x(2)
circ.h(range(3))
circ.ccx(0, 1, 2)
circ.h(range(2))
circ.x(range(2))
circ.h(1)
circ.cx(0, 1)
circ.h(1)
circ.x(range(2))
circ.h(range(2))
circ.measure_all()

arch = GenericBackendV2(
    num_qubits=5,
    coupling_map=[[0,1],[1,0],[1,2],[2,1],[1,3],[3,1],[3,4],[4,3]],
)

circ_comp = transpile(circ, backend = arch, optimization_level=1)

qcec.verify_compilation(circ, circ_comp)
```

 Documentation on all available configuration options is available at https://mqt.readthedocs.io/projects/qcec

 Details on the experimental setup, evaluations, and results can be found in [1].

References

1. L. Burgholzer, R. Raymond, R. Wille, Verifying results of the IBM Qiskit quantum circuit compilation flow, in *International Conference on Quantum Computing and Engineering* (2020). https://doi.org/10.1109/QCE49297.2020.00051
2. L. Burgholzer, R. Wille, Advanced equivalence checking for quantum circuits. IEEE Trans. CAD Integr. Circuits Syst. (2021). https://doi.org/10.1109/TCAD.2020.3032630
3. L. Burgholzer, R. Wille, QCEC: a JKQ tool for quantum circuit equivalence checking. Softw. Impacts (2021)

Despite the improvements made possible by the methods proposed in Chaps. 14 and 15, formal verification of quantum circuits may still reach its limits. But, as discussed above, simulation provides a promising alternative—motivating further investigation. In this regard, the simulation of quantum circuits on classical computer hardware is key (cf. Part II). Although this leads to exponential complexity in order to describe the corresponding quantum states and operations, powerful methods have recently been proposed to tackle this problem [1–8]. However, while the stimuli space for classical circuits is finite (each input bit can be assigned either 0 or 1—yielding a total of 2^n possible stimuli), the state space in the quantum realm is infinitely large (possible stimuli are elements of a 2^n-dimensional Hilbert space). This raises the question of whether simulative verification of quantum circuits (on classical computers) is suitable at all and, if so, how to generate proper stimuli to efficiently check the correctness of a quantum circuit. First promising results in that direction have already been demonstrated in Chap. 14, where it was shown that a couple of randomly chosen computational basis states already allow one to detect many cases of non-equivalence.

In this chapter (based on [9]), we show that although the perspective of a possible infinite number of stimuli may seem rather grim at first glance, there are promising ways to check the correctness of quantum circuits using simulative verification and random stimuli. However, this severely depends on how the stimuli are actually generated. In fact, we introduce, illustrate, and analyze three schemes for quantum stimuli generation, offering a nice trade-off between error detection rate (as well as the required number of stimuli) and efficiency. In contrast to classical circuits, we show (both theoretically and empirically) that even if only a few *randomly chosen* stimuli (generated from the proposed schemes) are considered, high error detection rates can be achieved in the quantum realm. The results of these conceptual and theoretical considerations have also been empirically confirmed, which, to the best of our knowledge, led to the broadest empirical evaluation of simulative verification schemes for quantum circuits to date—with a grand total of approximately 10^6 simulations conducted across 50 000 benchmark instances. Therefore, this chapter provides a more comprehensive assessment (both theoretically and empirically) of the "power of simulation" first proposed as part of a verification flow in Sect. 14.3.

The remainder of this chapter is structured as follows: Sect. 16.1 illustrates the general idea of performing simulative verification of quantum circuits. Then, Sect. 16.2 introduces, illustrates, and (theoretically) analyzes different stimuli generation schemes and their likelihood of detecting errors. The details of the implementation and a discussion of the experimental results are provided in Sect. 16.3.

16.1   General Idea

For the sake of this chapter, we consider the question whether a given circuit $G = g_0 \cdots g_{|G|-1}$ adheres to a given specification. Without loss of generality, we assume in the following that the specification is given as a unitary function U—possibly described by a high-level quantum algorithm, another circuit, or further functional representations for quantum computing. Due to the limited availability and accessibility of actual quantum computing devices and verification protocols that would require execution on these machines, such as [10, 11], we focus on efficient alternatives that can be employed prior to actual execution on a quantum computer (using classical computing devices). This has similarities to the verification of classical circuits, which shall also be conducted prior to an actual execution in the field.

A general scheme for simulative verification of quantum circuits has the following form:

1. Consider a set S of quantum states (which serve as stimuli).
2. Pick (and prepare) a quantum state $|\varphi\rangle \in S$.
3. Simulate (on a classical device) both U and G with this initial state—resulting in two states $|\varphi_U\rangle$ and $|\varphi_G\rangle$, respectively.
4. Compare the output $|\varphi_G\rangle$ generated by the realization G with the desired output $|\varphi_G\rangle$ by computing the quantum fidelity $\mathcal{F}$ between both states [12], i.e.,

$$\mathcal{F}(|\varphi_U\rangle, |\varphi_G\rangle) = |\langle\varphi_U|\varphi_G\rangle|^2 \in [0, 1].$$

In this regard, the fidelity $\mathcal{F}$ acts as a measure of similarity between two states—effectively computing the squared overlap of the states' amplitudes.

5. If $\mathcal{F}(|\varphi_U\rangle, |\varphi_G\rangle) \neq 1$, the stimulus $|\varphi\rangle$ shows the incorrect behavior of G with respect to U. Accordingly, the verification failed and the process is terminated.
6. Remove $|\varphi\rangle$ from S.
7. If $|S| \neq \emptyset$ (i.e., S is still non-empty) continue with Step (2); otherwise, the simulative verification process has been completed.

Now, the challenges of such an approach are as follows: First, to simulate a quantum circuit $G = g_0 \ldots g_{|G|-1}$ starting with an initial state $|\varphi\rangle$ on a classical device (Step (3) from above), matrix-vector multiplications of the matrices U_i (representing the circuit's gates g_i) with the state vector $|\varphi\rangle$ as well as the resulting output vectors, respectively, must be conducted consecutively. This leads to an exponential complexity, since the vectors and matrices involved have a size of 2^n and $2^n \times 2^n$, respectively (with n being the number of qubits). But although this is substantially harder than for the verification of classical circuits (here, a single simulation yields only linear complexity), rather powerful methods have been recently proposed to tackle this complexity—including methods based on highly optimized and parallel matrix-computations [1, 2], tensor networks [3, 4], quasiprobability/stabilizer-rank methods [5] (and references therein), as well as decision diagrams [6–8]. We refer to Part II for more information on the classical simulation of quantum circuits.

Second, as in the verification of classical circuits, the quality of the verification process is highly dependent on the applied set of stimuli, that is, 100% certainty cannot be guaranteed as long as the set of applied stimuli is not exhaustive. Furthermore, while the stimuli space for classical circuits is finite (each input bit can be assigned either 0 or 1—yielding a total of 2^n possible stimuli), the state space in the quantum realm is infinitely large (possible stimuli are elements of a 2^n-dimensional Hilbert space). This raises the question of whether simulative verification of quantum circuits (on classical computers) is suitable at all and, if so, how to generate proper stimuli $|\varphi\rangle$ to efficiently check the correctness of a quantum circuit.

In the following, we show that although the perspective of a possible infinite number of stimuli may seem rather grim at first glance, there are promising ways to check the correctness of quantum circuits using simulative verification. These, however, severely depend on how the stimuli are actually generated. In fact, we show (both theoretically and empirically) that high error detection rates can be achieved even if only a few randomly chosen stimuli are considered—as long as these are generated in a specific fashion.

16.2   Random Stimuli Generation

In this section, we propose different schemes for the generation of (random) stimuli and explore how well they can show the correctness of a quantum circuit. To this end, each of the following sections introduces, illustrates, and (theoretically) analyzes different stimuli generation schemes and their likelihood of detecting errors. Eventually, this will show that simulative verification indeed is very promising, since sets of stimuli can be generated in a fashion that offers a nice trade-off between error detection rate (as well as the required number of stimuli) and efficiency. The results of these conceptual and theoretical considerations have also been empirically confirmed as summarized later in Sect. 16.3.

16.2.1 Classical Stimuli

The most straightforward application of simulative verification for quantum circuits (compared to the classical approach) is to consider the set of computational basis states as stimuli (i.e., picking $|\varphi\rangle$ from the set $\{|i\rangle: \ i \in \{0, 1\}^n\}$) and computing $\mathcal{F}(U|i\rangle, V|i\rangle)$, where V is the matrix associated with G). This has been studied in [13, Sect. 8], where empirical results show that choosing "classical" stimuli from this set at random *often* allows to detect even small errors in quantum circuits. The following example illustrates this remarkable "power of simulation".

Example 16.1 Consider a certain n-qubit unitary specification U and assume that some error affects, without loss of generality, the first qubit in the actual realization G. In the quantum realm, this means that the circuit G is described by the unitary matrix $V = U \cdot (\mathbb{I}^{\otimes(n-1)} \otimes E)$, where E describes an error gate that is applied to the first qubit. Due to the inherent reversibility of quantum gates, this error has a localized effect on the output, i.e.,

$$\mathcal{F}(U|c\rangle, V|c\rangle) = \mathcal{F}(|c\rangle, (\mathbb{I}^{\otimes(n-1)} \otimes E)|c\rangle) = |\langle c_0|E|c_0\rangle|^2 \qquad (16.1)$$

for any classical stimulus $|c\rangle = |c_{n-1} \ldots c_0\rangle$. Now suppose that $E = X$, i.e., a bit flip error occurred. In contrast to classical intuition, such an error can be detected by a *single* simulation with *any* classical stimulus $|c\rangle$, since $\mathcal{F}(U|c\rangle, V|c\rangle) = |\langle c_0|X|c_0\rangle|^2 = 0$ independent of $|c\rangle$.

However, this approach has a severe handicap that has not been discussed so far—namely that it is not faithful. Specifically, for each unitary specification U there is an (infinitely large) family of realizations G for which $\mathcal{F}(U|c\rangle, V|c\rangle) = 1$ holds for all classical stimuli $|c\rangle$, even if quantum states $|\varphi\rangle$ with $\mathcal{F}(U|\varphi\rangle, V|\varphi\rangle) \neq 1$ actually exist. An example illustrates the problem:

Example 16.2 Consider the same scenario as in Example 16.1, but assume that the error is characterized as $E = Z$, i.e., a phase flip error occurred. No classical stimulus $|c\rangle$ can detect such an error due to the fact that $\mathcal{F}(U|c\rangle, V|c\rangle) = |\langle c_0|Z|c_0\rangle|^2 = 1$ independent of $|c\rangle$. Intuitively, this happens whenever the "difference" of U and V is diagonal in the computational basis, such as $\mathbb{I}^{\otimes(n-1)} \otimes Z$ in the case of this example. Note that such a difference may still be detected if alternative similarity measures between states are considered.

Nevertheless, empirical evaluations (which are summarized in Sect. 16.3) show that whenever classical stimuli are actually capable of detecting a certain error in the realization G, they do so within remarkably few simulations with randomly picked classical stimuli—an effect contradictory to classical intuition, as already observed in [14].

16.2.2 Local Quantum Stimuli

In the previous section, we showed that the generation of classical stimuli is not sufficient to faithfully detect errors in quantum circuits. On an abstract level, this should not come as a surprise. After all, quantum circuits are designed to achieve tasks that classical circuits cannot. In fact, a closer look at the single-(qu)bit case already reveals a fundamental discrepancy: Classical single-bit operations map one of two possible inputs (0 or 1) to one of two possible outputs (0 or 1). In contrast, the quantum case is much more expressive: The set of all possible single-qubit states $|\varphi\rangle$ is infinitely large and can be parametrized by the 2-dimensional Bloch sphere [12] illustrated in Fig. 16.1. Single-qubit quantum operations map single-qubit states to single-qubit states. Geometrically, this family encompasses all possible rotations of the Bloch sphere, as well as all reflections. Classical (single-qubit) stimuli, i.e., the states $|0\rangle$ and $|1\rangle$, are not expressive enough to reliably probe such a continuum of operations. They correspond to antipodal points on the (Bloch) sphere, and it is simply impossible to detect certain transformations by tracking the movement of only two antipodal points.

In order to address this, stimuli beyond (classical) basis states should also be considered. More precisely, three pairs of antipodal points are sufficient for full resolution [15–17], namely

$$|0\rangle, \qquad\qquad |1\rangle, \qquad\qquad (Z\text{-basis}), \qquad (16.2)$$
$$|+\rangle = 1/\sqrt{2}(|0\rangle + |1\rangle), \qquad |-\rangle = 1/\sqrt{2}(|0\rangle - |1\rangle), \qquad (X\text{-basis}), \text{ and} \quad (16.3)$$
$$|\uparrow\rangle = 1/\sqrt{2}(|0\rangle + i|1\rangle), \qquad |\downarrow\rangle = 1/\sqrt{2}(|0\rangle - i|1\rangle), \qquad (Y\text{-basis}). \quad (16.4)$$

Generating stimuli uniformly at random from this sextuple[1] produces a set that is expressive enough to detect *any* single-qubit error. More precisely, for any pair of functionally

Fig. 16.1 Bloch sphere

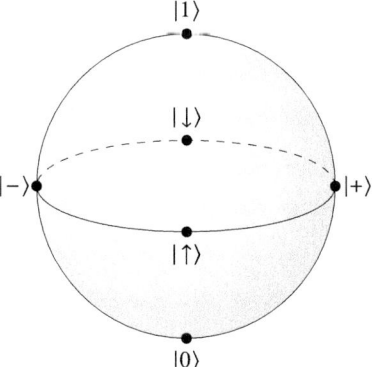

---

[1] The single-qubit states $|0\rangle, |1\rangle, |+\rangle, |-\rangle, |\uparrow\rangle, |\downarrow\rangle$ can be generated from the basis state $|0\rangle$ by applying the gates $\mathbb{I}, X, H, XH, HS,$ or XHS, respectively.

different single-qubit unitaries U and V, at least one input $|l_1\rangle \in \{|0\rangle, |1\rangle, |+\rangle, |-\rangle, |\uparrow\rangle, |\downarrow\rangle\}$ produces functionally different outputs, i.e., the fidelity $\mathcal{F}(U|l_1\rangle, V|l_1\rangle)$ is guaranteed to be $\neq 1$.

This desirable feature extends to the multi-qubit case. That is, if we independently select one of these six (single-qubit) states for every available qubit, every "local" single-qubit error may be detected. Thus, for n qubits, we consider the following ensemble of *local quantum stimuli*:

$$|l\rangle = |l_{n-1}\rangle \otimes \cdots \otimes |l_0\rangle \text{ with } |l_i\rangle \in \{|0\rangle, |1\rangle, |+\rangle, |-\rangle, |\uparrow\rangle, |\downarrow\rangle\} \tag{16.5}$$

Example 16.3 Recall the scenario from Example 16.1 (and Example 16.2). Compared to classical stimuli, local quantum stimuli behave in a more homogeneous fashion in the classical extreme cases shown before: First, suppose that $E = X$ (bit flip error). Then,

$$\mathcal{F}(U|l\rangle, V|l\rangle) = |\langle l_0|X|l_0\rangle|^2 = \begin{cases} 0 & |l_0\rangle \in \{|0\rangle, |1\rangle, |\uparrow\rangle, |\downarrow\rangle\} \\ 1 & |l_0\rangle \in \{|+\rangle, |-\rangle\} \end{cases}. \tag{16.6}$$

Compared to classical stimuli, only $2/3$ of all local quantum stimuli detect this type of error. Now, suppose that $E = Z$ (phase flip error). Then,

$$\mathcal{F}(U|l\rangle, V|l\rangle) = |\langle l_0|Z|l_0\rangle|^2 = \begin{cases} 0 & |l_0\rangle \in \{|+\rangle, |-\rangle, |\uparrow\rangle, |\downarrow\rangle\} \\ 1 & |l_0\rangle \in \{|0\rangle, |1\rangle\} \end{cases}. \tag{16.7}$$

Consequently, in contrast to not detecting such an error with classical stimuli at all, again $2/3$ of all local quantum stimuli are capable of detecting this type of error.

This observation that local quantum stimuli can detect errors which would have remained undetected using classical stimuli is not a coincidence. In fact, the collection of a total of 6^n local quantum stimuli is expressive enough to detect *any* error in a quantum circuit.

Theorem 16.1 *For each pair of functionally distinct n-qubit unitaries U and V, there exists at least one local quantum stimulus $|l\rangle$ as defined in Eq. (16.5) that detects the error, i.e., yields $\mathcal{F}(U|l\rangle, V|l\rangle) \neq 1$.*

Proof (Proof sketch) The key idea is to relate the expected fidelity $\mathbb{E}_{|l\rangle}\mathcal{F}(U|l\rangle, V|l\rangle)$—where the average is taken over all 6^n locally random stimuli—to a meaningful distance measure in the space of unitary matrices. This average outcome fidelity equals 1 if and only if U and V are functionally equivalent. Now, suppose that U and V are functionally distinct unitaries. Then, $\mathbb{E}_{|l\rangle}\mathcal{F}(U|l\rangle, V|l\rangle) < 1$ which is only possible if (at least) one stimulus $|l\rangle$ produces an outcome fidelity that is strictly smaller than one. $\qquad\square$

Although this rigorous statement asserts that any error can be detected by (at least) one local quantum stimulus, it does not provide any advice on how to find the "right" stimulus. This is a very challenging problem, in general, but the above example suggests that repeated random sampling of stimuli should "do the job". Our empirical studies (see Sect. 16.3) confirm that such a procedure works remarkably well. Typically, few randomly generated local quantum stimuli suffice to detect realistic errors.

16.2.3 Global Quantum Stimuli

The previous section has shown that a modest increase in the expressiveness of stimuli can already make a significant difference. Local quantum stimuli can detect *any* error, while classical stimuli cannot. This is interesting because local quantum stimuli are comparatively few in number (6^n states in a 2^n-dimensional state space to detect arbitrary discrepancies in unitary circuits) and actually do not inherit many further quantum features. For example, "global" quantum features such as entanglement are not employed by them at all. This begs the question: What kinds of advantages can even more expressive and "more quantum" stimuli offer? Faithfulness is no longer a problem, but richer global stimuli may help to detect errors *earlier*, i.e., after substantially fewer iterations.

In order to identify powerful global quantum stimuli, it is helpful to revisit local quantum stimuli as introduced in Eq. 16.5 from a different perspective. They are generated by starting with a very simple classical state (that is, $|0\ldots0\rangle$) and applying certain single-qubit gates to the individual qubits, e.g., $|0\rangle \otimes |+\rangle \otimes |\uparrow\rangle = (\mathbb{I} \otimes H \otimes HS)|000\rangle$. Consequently, *random* local stimuli are generated by choosing this *layer* of single-qubit gates at random. This generation scheme can be readily generalized. Rather than selecting only a single layer of (single-qubit) gates, we construct a generation circuit $G_0 \cdots G_{l-1}$ that has $l > 1$ layers and, most importantly, also features two-qubit gates. That is, a stimuli $|g\rangle$ with $|g\rangle = (G_0 \cdots G_{l-1})|0\ldots0\rangle$ is generated, where each G_i is a (single) layer comprised of so called Clifford gates (H, S, $CNOT$) [18]

In general, this set of *global quantum stimuli* $|g\rangle$ contains all local quantum stimuli but is much richer and much more expressive. For instance, the overwhelming majority of global quantum stimuli will be highly entangled. Provided that the number of layers l is proportional to the number of qubits n [19, 20], these stimuli show remarkable properties. Most notably, the expected outcome fidelity (averaged over all possible global quantum stimuli $|g\rangle$) accurately approximates one of the most prominent distance measures for n-qubit quantum circuits, namely

$$\mathbb{E}_{|g\rangle}\mathcal{F}(U|g\rangle, V|g\rangle) \approx \mathcal{F}_{\text{avg}}(U, V) = \tfrac{1}{2^n+1}\left(1 + 2^n\big|\text{tr}(U^\dagger V)\big|^2\right). \tag{16.8}$$

Here, $\text{tr}(U^\dagger V)$ denotes the trace of the unitary matrix $U^\dagger V$. This *average (gate) fidelity* [12] forms the basis of many state-of-the-art quantum calibration procedures [21, 22]. Importantly, most (realistic) errors lead to an average fidelity that is tiny. Equation 16.8 allows us to capitalize on this phenomenon. The following statement is an immediate consequence of Eq. 16.8 and Markov's inequality:

Corollary 16.1 *Consider a unitary specification U and a particular realization as a quantum circuit G (represented by the unitary V). Then, a randomly selected global quantum stimulus obeys*

$$\Pr_{|g\rangle} \left[\mathcal{F}(U|g\rangle, V|g\rangle) = 1 \right] \leq \mathcal{F}_{\mathrm{avg}}(U, V). \qquad (16.9)$$

The r.h.s. equals 1 if and only if G correctly realizes U, otherwise it is typically much smaller.

This general statement has powerful implications when applied to a precise example.

Example 16.4 Consider again the scenario from Example 16.1 (and Example 16.2): A single-qubit error E occurred on the first qubit leading to the unitary $V = U \cdot (\mathbb{I}^{\otimes(n-1)} \otimes E)$, where the single-qubit error is either $E = X$ (bit flip error) or $E = Z$ (phase flip error). Then, $\mathcal{F}_{\mathrm{avg}}(U, V) = \frac{1}{2^n+1} \leq 2^{-n}$ (because Pauli matrices are traceless) and Corollary 16.1 implies that it is very unlikely to *not* detect this error with a single random global quantum stimulus, i.e.,

$$\Pr_{|g\rangle} \left[\mathcal{F}(U|g\rangle, V|g\rangle) = 1 \right] \leq 2^{-n} \ll 1. \qquad (16.10)$$

This example demonstrates the power of global quantum stimuli. However, it is important to keep in mind that this power is not free. The generation of (random) global quantum stimuli and subsequent simulation is much more resource intensive by comparison (as confirmed by our empirical evaluations in Sect. 16.3).

This can also be understood from a broader context: The average (gate) fidelity as given by Eq. 16.8 is closely related to another popular distance measure—the *entanglement fidelity*. This quantity captures the performance of a powerful quantum stimulus $|\Omega\rangle$, see, e.g., [11]. This stimulus is generated from $2n$ qubits by pairwise entangling individual qubits of one half of the system with the qubits of the other half. Applying both circuits to the first half of this state and computing the fidelity of the outcome states subsequently yields the entanglement fidelity [23, 24], i.e.,

$$\mathcal{F}(U \otimes \mathbb{I}|\Omega\rangle, V \otimes \mathbb{I}|\Omega\rangle) = 4^{-n} \left| \mathrm{tr}(U^\dagger V) \right|^2 = \mathcal{F}_{\mathrm{ent}}(U, V). \qquad (16.11)$$

Comparing Eqs. 16.8 and 16.11 shows that these quantities are almost identical. This implies that global quantum stimuli accurately approximate the powerful quantum stimulus $|\Omega\rangle$ on average. Finally, we point out that conducting simulative verification with $|\Omega\rangle$ itself is not feasible on classical computers, since requiring twice the amount of qubits exponentially increases the resource demand for classical simulations.

16.3   Summary of Results

The schemes proposed in Sect. 16.2 have been implemented on top of the open-source MQT QCEC tool [25] introduced in Chap. 14. An empirical study on the behavior of the schemes proposed in Sect. 16.2 has been conducted in [9]—overall resulting in about one million simulations on 50 000 benchmark instances. To the best of our knowledge, this led to the broadest empirical evaluation of simulative verification schemes for quantum circuits to date. From those results, the following conclusions can be drawn:

- *All* schemes lead to sets of stimuli with remarkable error detection rates. With randomly chosen stimuli, few stimuli are sufficient to detect the vast majority of errors (whereas, in contrast, dedicated constrained-based stimuli generation or fuzzing methods [26–31] are required in the classical realm to obtain an acceptable error detection rate).
- Based on these high standards, classical stimuli generation performs worst and often fails—especially in cases where individual (diagonal) gates are removed or added. This is a consequence of the fact that classical stimuli are not faithful, as shown in Sect. 16.2.1. At the same time, the corresponding simulations are very fast; making this scheme suitable for rapid prototyping.
- On the other side of the spectrum, global stimuli generation produces the most robust results, i.e., requiring the least amount of stimuli and also reaches the highest error detection rates. This confirms the discussions of Sect. 16.2.3 on the quality of those stimuli. Thus, this scheme is suitable for rigorous testing even if the simulation of those stimuli is severely more runtime-demanding.
- The generation of local quantum stimuli constitutes a trade-off between quality and efficiency compared to the other two schemes. Although this scheme is not as powerful as global quantum stimuli generation in quality, it is faithful (as shown in Sect. 16.2.2) and remains rather efficient.

These results show that simulative verification offers huge potential in the verification of quantum circuits and that, in contrast to classical circuits, just generating a few randomly chosen stimuli according to the proposed schemes already achieves high error detection rates.

Implementation, Usage, Documentation, and Results

 The proposed stimuli generation schemes are available as part of the open-source MQT QCEC tool [25] at https://github . com/munich-quantum-toolkit/qcec. which can be installed using pip install mqt . qcec.

 The resulting tool can be setup as described and illustrated in section 14.5 on page 117. Then, to use the stimuli generation methods proposed in this chapter, only one line has to be modified: Instead of only calling

```
qcec.verify(qc1, qc2)
```

the verification function can be configured with

```
qcec.verify(qc1, qc2, state_type='local_random',
max_sims=5)
```
to run a maximum of five simulations with random local quantum stimuli.

 Documentation on all available configuration options is available at https : //mqt . readthedocs . io/projects/qcec

 Details on the experimental setup, evaluations, and results can be found in [9].

References

1. G.G. Guerreschi, J. Hogaboam, F. Baruffa, N.P.D. Sawaya, Intel quantum simulator: a cloud-ready high-performance simulator of quantum circuits. Quant. Sci. Technol. **5**(3), 034–007 (2020). ISSN: 2058-9565. https://doi.org/10.1088/2058-9565/ab8505
2. T. Jones, A. Brown, I. Bush, S.C. Benjamin, QuEST and high performance simulation of quantum computers. Sci. Rep. (2018)
3. B. Villalonga, S. Boixo, B. Nelson, et al., A flexible high-performance simulator for verifying and benchmarking quantum circuits implemented on real hardware. NPJ Quant. Inf. (2019). ISSN: 2056-6387. https://doi.org/10.1038/s41534-019-0196-1
4. E. Pednault, J.A. Gunnels, G. Nannicini, L. Horesh, R. Wisnieff, Leveraging secondary storage to simulate deep 54-qubit Sycamore circuits. arXiv: 1910.09534 (2019), preprint
5. J.R. Seddon, B. Regula, H. Pashayan, Y. Ouyang, E.T. Campbell, Quantifying quantum speedups: improved classical simulation from tighter magic monotones. arXiv: 2002.06181 (2020), preprint
6. P. Niemann, R. Wille, D.M. Miller, M.A. Thornton, R. Drechsler, QMDDs: efficient quantum function representation and manipulation. IEEE Trans. CAD Integr. Circuits Syst. (2016)

7. A. Zulehner, R. Wille, Advanced simulation of quantum computations. IEEE Trans. CAD Integr. Circuits Syst. (2019). https://doi.org/10.1109/TCAD.2018.2834427

8. A. Zulehner, R. Wille, Matrix-Vector vs. Matrix-Matrix multiplication: potential in DD-based simulation of quantum computations, in *Design, Automation and Test in Europe* (2019). https://doi.org/10.23919/DATE.2019.8714836

9. L. Burgholzer, R. Kueng, R. Wille, Random stimuli generation for the verification of quantum circuits, in *Asia and South Pacific Design Automation Conference* (2021). https://doi.org/10.1145/3394885.3431590

10. J. Watrous, *The Theory of Quantum Information* (Cambridge University Press, 2018)

11. S. Khatri, R. LaRose, A. Poremba, L. Cincio, A.T. Sornborger, P.J. Coles, Quantum-assisted quantum compiling. Quantum **3**, 140 (2019)

12. M. A. Nielsen and I. L. Chuang, *Quantum Computation and Quantum Information* (Cambridge University Press, 2010)

13. L. Burgholzer, R. Wille, Advanced equivalence checking for quantum circuits. IEEE Trans. CAD Integr. Circuits Syst. (2021). https://doi.org/10.1109/TCAD.2020.3032630

14. L. Burgholzer, R. Wille, The power of simulation for equivalence checking in quantum computing, in *Design Automation Conference* (2020)

15. J. Schwinger, Unitary operator bases. Proc. Natl. Acad. Sci **46**(4), 570–579 (1960)

16. A. Klappenecker, M. Rotteler, Mutually unbiased bases are complex projective 2-designs, in *International Symposium on Information Theory* (2005), pp. 1740–1744

17. R. Kueng, D. Gross, Qubit stabilizer states are complex projective 3-designs. arXiv: 1510.02767 (2015), preprint

18. D. Gottesman, Stabilizer codes and quantum error correction. Caltech (1997)

19. N. Hunter-Jones, Unitary designs from statistical mechanics in random quantum circuits. arXiv: 1905.12053 (2019), preprint

20. F.G.S.L. Brandão, A.W. Harrow, M. Horodecki, Local random quantum circuits are approximate polynomial-designs. Commun. Math. Phys. (2016)

21. E. Magesan, J.M. Gambetta, J. Emerson, Characterizing quantum gates via randomized benchmarking. Phys. Rev. A **85**(4), 042–311 (2012)

22. R. Kueng, D.M. Long, A.C. Doherty, S.T. Flammia, Comparing experiments to the fault-tolerance threshold. Phys. Rev. Lett. **117**(17), 170–502 (2016)

23. B. Schumacher, Sending entanglement through noisy quantum channels. Phys. Rev. A **54**(4), 2614–2628 (1996). https://doi.org/10.1103/PhysRevA.54.2614

24. A. Kitaev, Quantum computations: algorithms and error correction. Russ. Math. Surv. **52**(6), 1191–1249 (1997)

25. L. Burgholzer, R. Wille, QCEC: a JKQ tool for quantum circuit equivalence checking. Softw. Impacts (2021)

26. J. Yuan, C. Pixley, A. Aziz, *Constraint-Based Verification* (Springer, 2006)

27. J. Bergeron, *Writing Testbenches Using System Verilog* (Springer, 2006)

28. N. Kitchen, A. Kuehlmann, Stimulus generation for constrained random simulation, in *International Conference on CAD* (2007), pp. 258–265

29. R. Wille, D. GroSSe, F. Haedicke, R. Drechsler, SMT-based stimuli generation in the SystemC Verification library, in *Forum on Specification and Design Languages* (2009)

30. H.M. Le, D. GroSSe, N. Bruns, R. Drechsler, Detection of hardware Trojans in SystemC HLS designs via coverage-guided fuzzing, in *Design, Automation and Test in Europe* (2019)

31. K. Laeufer, J. Koenig, D. Kim, J. Bachrach, K. Sen, RFUZZ: coverage-directed fuzz testing of RTL on FPGAs, in *International Conference on CAD* (2018)

Equivalence Checking of Quantum Circuits with the ZX-Calculus

In the literature, the *ZX*-calculus [1–4] (cf. Sect. 3.3) has been mainly used for compilation and optimization of quantum circuits. However, in [1], it is briefly mentioned that the optimization algorithm proposed in that paper can also be used to check the equivalence of two quantum circuits. The general idea is similar to the methodology proposed in Chap. 14. One of the circuits is inverted and combined with the remaining circuit. Then, the rules of the ZX-calculus are used to simplify the resulting diagram as much as possible. If, at the end, a diagram results that only consists of bare wires, the circuits can be considered equivalent (up to some permutation of the qubits). Unfortunately, the method was not explicitly designed for equivalence checking, and, as a result, does not address many of the issues unique to the problem, such as inaccurate representations of complex numbers, different (circuit-to-device) qubit mappings, and ancillary qubits. Furthermore, the only existing implementation of this method (PyZX, Kissinger and van de Wetering [5]) is written in Python, which is known to yield suboptimal performance compared to compiled languages like C++. Apart from practical considerations, theoretical aspects have also hardly been investigated so far. It is not known for which class of circuits the equivalence checking method based on the ZX-calculus is complete, i.e., whether it can actually prove the equivalence of any two equivalent quantum circuits.

Motivated by this, the aim of this chapter (based on [6, 7]) is twofold. First, to establish whether the ZX-calculus provides a solid equivalence checking methodology, we review the current state of the art in equivalence checking with the ZX-calculus. To expand on this, we discuss how this method can be augmented to handle inaccurate representations of complex numbers arising from compilation and optimization processes, deal with alterations of the input and output layouts of a circuit that occur during compilation, and integrate ancillary qubits into the equivalence checking procedure. We provide the first results on the completeness of this equivalence checking algorithm. Secondly, to empirically show that the

© The Author(s), under exclusive license to Springer Nature Switzerland AG 2026
L. Burgholzer and R. Wille, *Design Automation Tools and Software for Quantum Computing*, https://doi.org/10.1007/978-3-032-06770-8_17

ZX-calculus is a practically relevant method in equivalence checking, we conduct a detailed case study to establish a baseline for the current state of the art in equivalence checking of quantum circuits, considering a wide range of benchmarks.

The remainder of this chapter is structured as follows: Sect. 17.1 explains the state-of-the-art equivalence checking algorithm based on the ZX-calculus. Then, Sect. 17.2 shows how the method can be expanded to handle more relevant equivalence checking problems in the compilation and optimization of quantum circuits. In Sect. 17.3, we prove that the equivalence checking method based on the ZX-calculus is complete for Clifford circuits and illustrate that it is not complete, in general. Section 17.4 concludes with a discussion of the resulting implementation and a summary of the experimental results.

17.1 Equivalence Checking of Graph-Like ZX-Diagrams

Equivalence checking with the ZX-calculus can be done in one of two ways, by either rewriting the diagrams of both circuits into one another or by inverting one diagram, composing the diagrams, and simplifying as much as possible. This is sometimes called an *equivalence checking miter*. If the composed diagram simplifies to a diagram composed only of bare wires, it is either the identity or contains swaps, i.e., it resembles a permutation.

Example 17.1 Consider the circuit G

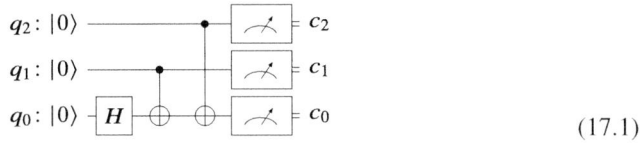

$$(17.1)$$

and assume it has been mapped to a linear five-qubit architecture as follows:

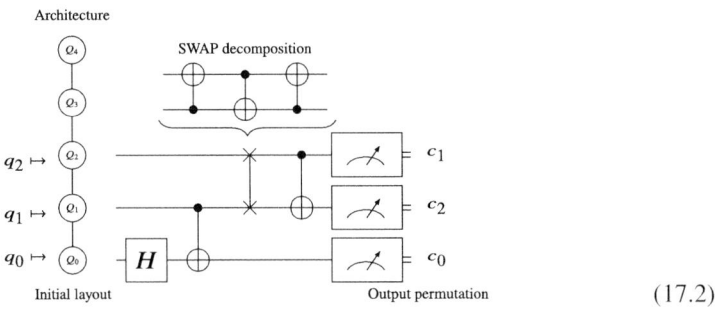

$$(17.2)$$

Note that, initially, qubit q_i of the circuit is mapped to qubit Q_i on the device for $0 \leq i \leq 2$ and that, at the end of the circuit, q_0 is measured on Q_0, q_1 on Q_2 and q_2 on Q_1.

Then, the ZX-diagrams of both circuits are given by:

$$(17.3)$$

Since all phases in all spiders are 0, the inverse of each diagram is obtained by just reversing the diagram. Using the rewrite rules of the ZX-calculus to prove the identity of the circuits proceeds as follows:

$$(17.4)$$

$$(17.5)$$

The diagram contains a SWAP which permutes qubit Q_1 and Q_2. Since this is what is expected from the output permutation described above, it can be concluded that the circuits are equivalent.

This example shows that the ZX-calculus cannot only show the equivalence of circuits, but that it can also provide a proof certificate in the form of the order of rewrite rules that are applied to derive the identity. However, the standard set of equations (cf. Sect. 3.3) is not suitable for automated equivalence checking. For example, in Eq. 3.31, the spider fusion rule is used in both directions, which is undesirable in automated rewriting, where *terminating* rewriting systems are preferable.

In [3], the authors introduce an alternative structure for ZX-diagrams coupled with additional rewrite rules. A ZX-diagram is *graph-like* when:

1. All spiders are Z-spiders.
2. Z-spiders are only connected via Hadamard edges.
3. There are no parallel Hadamard edges or self-loops.
4. Every input or output is connected to a Z-spider, and every Z-spider is connected to at most one input or output.

In graph-like ZX-diagrams, a spider connected to an input or output is called a *boundary* spider. Otherwise, it is called an *interior* spider. Most importantly, every ZX-diagram is equal to a graph-like ZX-diagram [3]. Every ZX-diagram can be rewritten to its equivalent graph-like form using the basic rules given in Fig. 3.1. Instead of rigorously defining this rewriting procedure, we give an intuition with the following example.

Example 17.2 The ZX-diagram of the compiled GHZ state preparation circuit from Example 17.1 can easily be transformed to its equivalent graph-like form by applying the rules (**id**) and (**h**).

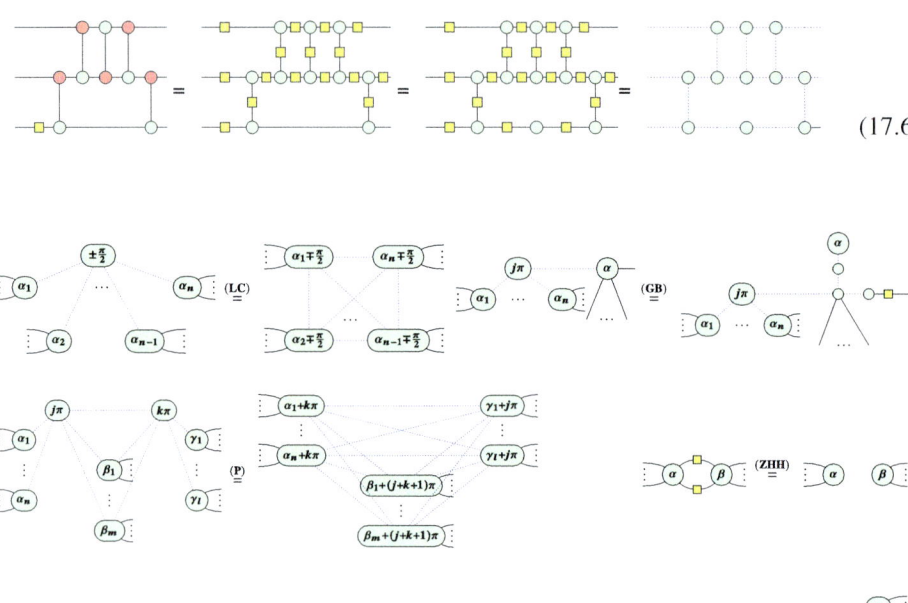

$$(17.6)$$

Fig. 17.1 Rewrite system for graph-like ZX-diagrams

Graph-like diagrams allow for the formulation of a normalizing rewrite system. The rules of this system are given in Fig. 17.1. Adding rules (**f**), (**id**) and (**hh**) (applied from left to right) to this system allows for the definition of a simplification algorithm that reduces every ZX-diagram into a *reduced gadget form* [4], which, in turn, allows for the definition of an equivalence checking algorithm: Given two quantum circuits G and G', their equivalence can be checked by taking their respective representations as ZX-diagrams D and D', combining them to $D^\dagger D'$, and simplifying the combined diagram to reduced gadget form. If the reduced gadget form is the identity diagram—the ZX-diagram consisting only of bare wires—then G and G' are equivalent. Otherwise, nothing can be concluded about the relation of G and $G^\dagger$ because, in general, there is no unique reduced gadget form for a ZX-diagram.

17.2   Adaptations for Equivalence Checking

The original ZX-calculus equivalence checking algorithm proposed in [1] has been introduced as a by-product of the optimization algorithm proposed in that work. It has, therefore, not been expanded to handle the more technical aspects of quantum circuit equivalence checking necessary to check the results of compilation flows. In the following, we will remedy this by showing how inaccuracies, permutations, and ancilla qubits can be handled in equivalence checking using the ZX-calculus.

17.2.1 Handling Inaccuracies

The equivalence checking routine based on the ZX-calculus is an exact method. Therefore, when considering two quantum circuits G and G', where G' is equivalent to G up to some small error, the ZX-calculus cannot conclude the equivalence. But can anything be concluded about the reduced gadget form of $D^\dagger D'$ where $[\![D]\!] \approx [\![D']\!]$?

Example 17.3 To give an intuition, consider the Clifford+T circuit

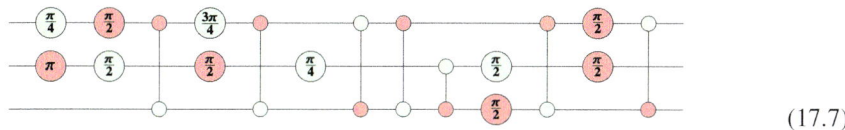

$$(17.7)$$

Introducing an error of 10^{-15} in the phase of two spiders and checking the equivalence of the original and the erroneous circuit yields a ZX-diagram that looks like:

$$(17.8)$$

The phases indicated with $\sim \alpha$ means that the phase is $\alpha \pm 10^{-15}$. It is not at all obvious that this diagram is close to the identity. However, if the phases were rounded and the diagram was simplified further, the identity would indeed be derived.

One strategy to overcome this limitation is to interleave normal-form simplification with detection and rounding of phases close to $k\frac{\pi}{2}$ for some $k \in \mathbb{Z}$. This allows the equivalence checking of quantum circuits that differ by small numerical inaccuracies in some continuous parameters. The corresponding algorithm is obviously not correct in a formal sense, i.e., it can attest two non-equivalent circuits to be equivalent, but that is the whole point. The threshold for rounding ϵ can be used to tune the degree to which errors are allowed. However, it does not give any indication about the absolute error. To clarify this point, consider a ZX-diagram M after simplification and the corresponding ZX-diagram M' obtained after rounding and simplifying again. The tolerance ϵ cannot be used to assess $\mathrm{tr}([\![D^\dagger]\!][\![D']\!])$—the Hilbert-Schmidt inner product discussed in Chap. 13. A question one might ask is why the rounding does not already occur on the diagrams D and D'. The reason is that even phases that are not nice (or very small) fractions of π might cancel during simplification due to the rules

UG and **GF**. Thus, rounding before simplifying would increase the total error made during the equivalence check.

This way of handling inaccuracies is still lacking. As discussed above, it is hard to gauge the tolerance required in order to ensure that the absolute error allowed is within some bound. Given the ZX-diagram M after full simplification, how can it be determined whether $|\operatorname{tr}([\![M]\!])| \approx 2^n$? The diagrammatic trace is defined as follows:

$$\operatorname{tr}\left(\; \vdots \; \boxed{D} \; \vdots \; \right) = \left(\vdots \; \boxed{D} \; \vdots \right) \tag{17.9}$$

Unfortunately, this definition is hardly helpful to actually compute the trace, which requires further simplifications after the inputs and outputs have been connected. This does not necessarily enable the ZX-diagram to be simplified to a point where calculations are practical. A possible solution to this problem is to leave the ZX-calculus framework entirely. ZX-diagrams are, in essence, tensor networks [8, 9]. Therefore, methods from the tensor network domain can be used to compute the trace of a ZX-diagram.

17.2.2 Handling Permutations

Handling SWAPs in ZX-diagrams is a trivial matter. Since SWAPs are nothing but edges connecting spiders acting on different qubits, they do not add much complexity to a ZX-diagram. To correct permutations of the initial layout, it has to be ensured that the wires are connected accordingly when building $D^\dagger D'$. But since SWAPs incur such little overhead in ZX-diagrams, the initial layout can also just be encoded into the original diagrams themselves before performing the equivalence check. The wires can then be connected in the usual way, i.e., by connecting the ith output wire of $D^\dagger$ with the ith input wire of D'. Output permutations can be handled in a similar way as with decision diagrams, by comparing the permutation of wires after fully simplifying $D^\dagger D'$ with the expected permutation. Once again the permutation can also be handled by encoding the SWAPs directly into the diagrams. When converting a compiled circuit to a ZX-diagram, Eq. 3.32 can be used to reconstruct compiled SWAP gates if they are not optimized away (similar to the optimization shown in Subsect. 15.2.3). Since a SWAP in the ZX-calculus is only a crossing of the wires, this reconstruction can greatly improve the performance of the equivalence check, decreasing runtime by up to two orders of magnitude.

Example 17.4 The output permutation of the compiled circuit in Example 17.1 can be directly encoded back into the circuit via a SWAP at the end of the circuit. Additionally, the 3 CNOTs can be converted back into a SWAP, producing the following ZX-diagram:

$$(17.10)$$

This ZX-diagram is equivalent to the diagram of the original circuit from Example 17.1 up to "untangling" the wires.

17.2.3 Handling Ancillaries

In the ZX-calculus, the equivalence checking problem using ancillaries is reducible to the ancilla-free case in a straightforward fashion. Remember that ancillaries are qubits that have a constant initial state and end in the same state. This is easily translated into the diagrammatic language of the ZX-calculus, by replacing each input and output belonging to an ancilla qubit by the respective state and effect, which are just X-spiders with a phase of either 0 or π.

Example 17.5 It can be shown in the ZX-calculus that fixing the control qubit of a CNOT gate to the $|1\rangle$ state, transforms it into a Pauli X gate, by only basic applications of the axioms:

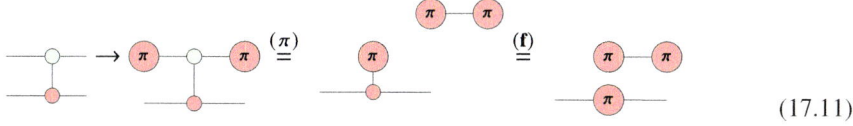

$$(17.11)$$

Since scalars can be ignored, the right-hand side indeed implements a Pauli X gate ( ).

17.3 Completeness

A natural question to ask is whether the ZX-calculus is powerful enough to derive the identity for any pair of functionally equivalent circuits. The good news is that the ruleset provided in this paper is complete for circuits solely composed of Clifford gates [10]. As such, it is not surprising that automated equivalence checking with the ZX-calculus is also complete for Clifford ZX-diagrams. The bad news is that to achieve completeness for universal quantum computing, the ruleset has to be extended with a rule involving complicated iterated trigonometric functions [11], making it difficult to apply in automated reasoning.

Theorem 17.1 *Given two equivalent quantum circuits G and G′ consisting only of Clifford gates with corresponding ZX-diagrams D and D′ the only reduced gadget form of $D^\dagger D′$ is the identity diagram.*

Proof See [6].   □

Theorem 17.1 establishes a baseline for what equivalences can be proven through automated reasoning with the ZX-calculus. Next, we want to look at completeness from a different perspective, by showing that rewriting to reduced gadget form is not sufficient for proving the equivalence of arbitrary equivalent circuits. In particular, we show that this algorithm is not even sufficient for proving the equivalence of reversible circuits.

Theorem 17.2 *There exist two equivalent quantum Circuits G and G′—using ancillary qubits—with corresponding ZX-diagrams D and D′ where $D^\dagger D′$ possesses a reduced gadget form that is not the identity diagram.*

Proof The proof of this theorem is not achieved through cunning manipulation of diagrams, kets, and bras, but by brute-force calculation. Consider the following ZX-diagram for a multi-controlled Toffoli gate (requiring no additional qubits):

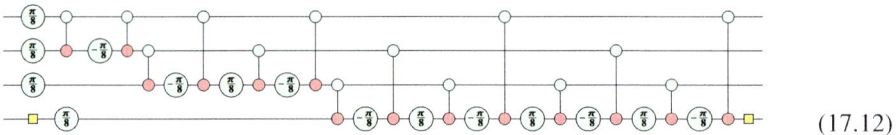

$$(17.12)$$

Also consider an alternative version using a single ancilla:

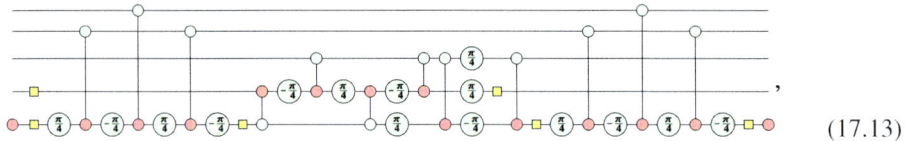

$$(17.13)$$

where the additional qubit's input and output has been set to $|0\rangle = $ ●—.

Taking the adjoint of the first diagram and concatenating the two diagrams yields:

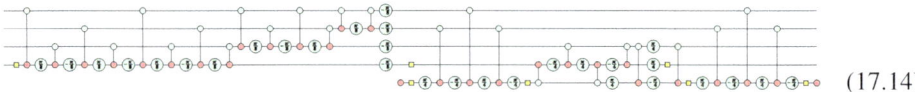

$$(17.14)$$

A reduced gadget form of this diagrams is given by:

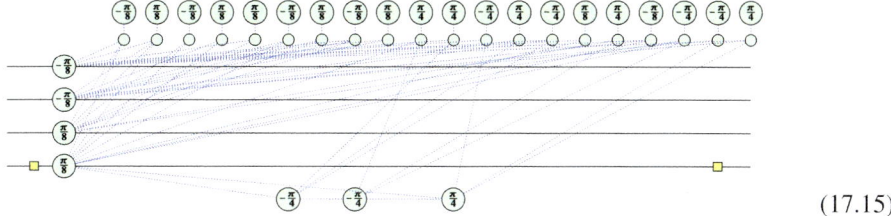

$$(17.15)$$

No further simplifications can be made but this diagram is clearly not the identity. However, it can be checked by (tedious) computation of the corresponding matrices that the above ZX-diagram actually implements the identity transformation. □

It is not surprising at all that equivalence checking via simplification to reduced gadget form is not complete, in general. Because the equivalence checking problem is QMA-complete and since NP is a subset of QMA, it would be completely unexpected that the ZX-calculus-based equivalence checking algorithm solves the equivalence checking problem, given that a reduced gadget form can be derived in polynomial time with respect to the number of spiders of the original diagram [4]. But the proof by counterexample shows that it cannot even show the equivalence of two fairly simple circuits (involving ancillaries).

17.4 Summary of Results

The equivalence checking method proposed in this chapter has been implemented on top of the open-source MQT QCEC tool [12] introduced in Chap. 14. An extensive case study has been conducted to evaluate the resulting implementation of the ZX-calculus-based equivalence checking algorithm against two existing equivalence checking tools—an approach based on path sums [13] and the DD-based methodology proposed in Chap. 14 that is available in QCEC. These evaluations have shown:

- The proposed method exhibits the same asymptotic scaling as the path-sum tool Feyn-Ver [13] on a set of random Clifford benchmarks. However, in absolute numbers, it outperforms Feynver by orders of magnitude on all benchmarks.
- Although this is not guaranteed by the theory of the ZX-calculus, the proposed method managed to prove the correct result for all circuits considered. This suggests that the ZX-checker failing to prove the equivalence of two circuits might provide an indication, although no guarantee, that the considered circuits are not equivalent.
- For circuits with many qubits but fewer gates, the ZX-calculus-based approach performs more favorably compared to the methodology proposed in Chap. 14.

- For larger circuits, the problem itself is too complex to be solved for the ZX-calculus-based approach. In this case, the methodology proposed in Chap. 14 might still be able to show equivalence by keeping the intermediate representations compact.

Overall, we conclude that neither the ZX-calculus nor the decision diagram-based method is clearly better than the other, but that they are best used in tandem, especially for more optimized circuits. Because the size of the ZX-diagram during the equivalence check is bounded by the size of the original circuit (the number of spiders is strictly decreasing), the ZX-calculus-based method has a low memory footprint. Therefore, it can easily be used in parallel with the decision diagram-based method without using too many resources.

Implementation, Usage, Documentation, and Results

 The resulting equivalence checker based on the ZX-calculus is available as part of the open-source MQT QCEC tool [12] at
https://github . com/munich-quantum-toolkit/qcec,
which can be installed using pip install mqt . qcec.

 The resulting tool can be setup as described and illustrated in section 14.5 on page 117. Using the ZX-checker requires no modifications to the setup, as it is enabled per default in QCEC. In the parallel mode, the checker is started concurrently with the DD-based checkers. In sequential mode, it is started after all DD-based checkers.

 Documentation on all available configuration options is available at
https://mqt . readthedocs . io/projects/qcec

 Details on the experimental setup, evaluations, and results can be found in [6], [7].

References

1. J. van de Wetering, ZX-calculus for the working quantum computer scientist (2020). arXiv preprint: arXiv:2012.13966
2. A. Cowtan, W. Simmons, R. Duncan, A generic compilation strategy for the unitary coupled cluster ansatz (2020). arXiv preprint: arXiv:2007.10515
3. R. Duncan, A. Kissinger, S. Perdrix, J. van de Wetering, Graph-theoretic simplification of quantum circuits with the ZX-calculus (2019). arXiv preprint: arXiv:1902.03178
4. A. Kissinger, J. van de Wetering, Reducing T-count with the ZX-calculus. Phys. Rev. A (2020)
5. A. Kissinger, J. van de Wetering, PyZX: large scale automated diagrammatic reasoning, **318**, 229–241 (2019)
6. T. Peham, L. Burgholzer, R. Wille, Equivalence checking of quantum circuits with the ZX-calculus. jetcas (2022). https://doi.org/10.1109/JETCAS.2022.3202204
7. T. Peham, L. Burgholzer, R. Wille, Equivalence checking paradigms in quantum circuit design: a case study, in *Design Automation Conference* (2022)
8. B. Coecke, A. Kissinger, Picturing quantum processes, in *Diagrammatic Representation and Inference*, ed. by P. Chapman, G. Stapleton, A. Moktefi, S. Perez-Kriz, F. Bellucci (2018)
9. J.D. Biamonte, V. Bergholm, Tensor networks in a nutshell (2017). arXiv preprint: arXiv:1708.00006
10. M. Backens, The ZX-calculus is complete for stabilizer quantum mechanics (2013). arXiv preprint: arXiv:1307.7025
11. R. Vilmart, A near-optimal axiomatisation of ZX-calculus for pure qubit quantum mechanics (2018). arXiv preprint: arXiv:1812.09114
12. L. Burgholzer, R. Wille, QCEC: a JKQ tool for quantum circuit equivalence checking. Software Impacts (2021)
13. M. Amy, Towards large-scale functional verification of universal quantum circuits, in *International Conference on Quantum Physics and Logic* (2019)

The capabilities of quantum computers built today are rapidly increasing. New devices not only feature more and more qubits, which are less prone to errors, but also allow for a much tighter classical control loop. This is witnessed by the OpenQASM 3.0 specification published by IBM [1] and the ability to perform conditional resets on IBM quantum computers [2]. Through the interaction of classical computation with the gates and measurements of a quantum circuit, new computing primitives such as mid-circuit measurements and resets as well as classically controlled operations become possible within the coherence time for a single circuit execution. We use IBM's terminology and refer to this new, larger class of circuits as *dynamic quantum circuits*. With these rapid advances in physical realizations comes the need for quantum software that aids developers and users in keeping up with this pace. Besides posing challenges, e.g., for classical and quantum design and compilation in general, this also poses new challenges for quantum circuit verification. Almost all known approaches to addressing this problem (including those proposed in the previous chapters) assume that the underlying functionality is unitary, which dynamic circuit primitives are not. Therefore, it is not possible to use established techniques for verifying conventional quantum circuits in a straightforward way.

In this chapter (based on [3]), we discuss the effects of non-unitaries on quantum circuit equivalence checking. We also show that it is not necessary to start from scratch in order to use existing tools to handle this broader class of circuits. To this end, we propose two different schemes that target two slightly different verification scenarios. First, we ask if two circuits, G and G', which may use dynamic circuit primitives, are functionally equivalent as a whole. We show that any such circuit can be *transformed* into a circuit with only unitary operations by combining well-known results from quantum information. By changing the primitives of dynamic circuits in this way, existing methods for checking the equivalence of quantum circuits can be used for this broader class of dynamic circuits. Second, we consider

© The Author(s), under exclusive license to Springer Nature Switzerland AG 2026
L. Burgholzer and R. Wille, *Design Automation Tools and Software for Quantum Computing*, https://doi.org/10.1007/978-3-032-06770-8_18

the question of whether two circuits G and G' produce the same distribution of measurement outcomes given a fixed input state, i.e., whether they behave the same when executed on a quantum computer. We show how to use classical quantum circuit simulation in a clever way to get the full measurement probabilities of a dynamic circuit as if it did not contain any non-unitaries. Overall, these schemes form a generic solution for handling non-unitaries in verifying the equivalence of quantum circuits that is applicable to any existing verification framework.

The remainder of this chapter is structured as follows: Sect. 18.1 introduces dynamic circuits and explains the resulting problem of equivalence checking in detail, along with the general idea for solving this problem. Then, Sects. 18.2 and 18.3 elaborate on the proposed schemes and provide some discussion. section 18.4 concludes with a discussion of the resulting implementation and experimental results.

18.1 Dynamic Circuits and Resulting Problem

The circuit model of quantum computing has been the de-facto standard for designing quantum circuits to be executed on current generation quantum computers. However, this describes quantum circuits in a *static* fashion—with no opportunity to steer the computation in a direction based on outcomes of intermediate results. Recently, IBM announced that their quantum computers now allow for interactions with classical computing instructions within the runtime of a quantum circuit—enabling what IBM refers to as *dynamic quantum circuits* [2]. In the following, we describe what constitutes these new kinds of circuits and discuss the resulting challenges for checking the equivalence of quantum circuits that might contain dynamic circuit primitives. Before that, we introduce a running example that will be used throughout the rest of this chapter to illustrate the concepts.

18.1.1 Running Example

The key ideas of this chapter will be illustrated by means of a particular quantum algorithm, namely *Quantum Phase Estimation* (QPE, Nielsen and Chuang [4]), which represents one of the key subroutines in important quantum algorithms such as Shor's algorithm [5] for factoring numbers, the HHL algorithm [6] for solving linear systems, or quantum principal component analysis [7] for machine learning. It solves the problem of determining the phase of a unitary operator U given an eigenstate $|\psi\rangle$, that is, determining $\theta \in [0, 1)$ such that $U |\psi\rangle = e^{2\pi i \theta} |\psi\rangle$. To this end, the QPE algorithm determines an m-bit estimate $\tilde{\theta} = 0.c_{m-1} \ldots c_0$ of θ. First, controlled-U^{2^k} operations ($0 \leq k < m$) are used to write the m-bit Fourier basis representation of U's phase to an m-qubit register. Afterward, the inverse *Quantum Fourier Transform* (QFT†, Nielsen and Chuang [4]) is applied to transform the result into the computational basis. Whenever θ is representable using m fractional bits, the algorithm succeeds with certainty, while otherwise it gives a suitably high chance of success (with a probability larger than $\frac{4}{\pi^2} \approx 0.405$).

Example 18.1 Assume U is given by $p(\frac{3\pi}{8}) = \text{diag}(1, e^{2\pi i \frac{3}{16}})$ and $|\psi\rangle = |1\rangle$. Then, the quantum circuit realizing the 3-bit precision QPE algorithm has the following form:

$$\tag{18.1}$$

It applies three rounds of controlled-phase rotations and then uses the three-qubit inverse Fourier transform to obtain the desired estimate $\tilde{\theta} = 0.c_2 c_1 c_0$ from the measurement results. Since $\theta = \frac{3}{16} = 0.0011_2$ cannot be exactly represented using three fractional bits, running the algorithm yields $|001\rangle$ and $|010\rangle$ as the most probable output states.

18.1.2 Dynamic Quantum Circuits and Their Benefits

By allowing the interaction of real-time classical computations with the gates and measurements of traditional quantum circuits, the quantum circuit model is extended by non-unitary primitives such as mid-circuit measurements and resets as well as classically controlled quantum operations. As a consequence, the circuits are no longer static, but rather *dynamic*. Eventually, these primitives will be necessary for quantum computers to achieve fault tolerance by realizing quantum error correction schemes. However, already in the near term, interesting use cases for teleportation [8] and algorithms like *Iterative QPE* (IQPE, Dobsicek et al. [9]) arise that employ dynamic circuit primitives in order to, e.g., reduce the required number of qubits—a limited resource thus far.

For example, our running example, i.e., the QPE algorithm reviewed in Sect. 18.1.1, may exploit non-unitaries to reduce the number of qubits: Instead of an m-qubit register for computing the Fourier base representation of the unitary's phase, a single qubit is used and repeatedly measured. Starting from the least significant bit of the resulting estimate $\tilde{\theta} = 0.c_{m-1} \ldots c_0$, each measurement adds one bit of information to the estimated phase. The result of each measurement then influences the rotation angles applied to the working qubit in the next iteration. This requires the availability of the measurement results and application of quantum operations based on them within the coherence time of the quantum computer's qubits. Researchers from IBM Quantum have recently demonstrated one of the first realizations of the IQPE algorithm in an actual system on one of their devices [10].

Example 18.2 Assume again that, as in Example 18.1, we want to estimate the phase θ of the unitary operator $U = p(\frac{3\pi}{8})$ corresponding to the eigenvector state $|\psi\rangle = |1\rangle$ up to a precision of three bits. Then, an alternative realization using dynamic circuit primitives is given by:

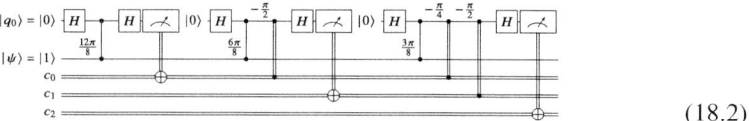

$$(18.2)$$

Instead of the 3-qubit register considered before in Example 18.1, a single working qubit in combination with mid-circuit measurements, resets, and classically controlled single-qubit rotations is used to iteratively compute individual bits of the phase estimate. Compiling this circuit to an actual device requires no mapping at all, since only two qubits interact with each other. As a consequence, the quantum cost of the resulting circuit is considerably reduced—significantly improving the expected fidelity when executing the circuit on an actual device.

18.1.3 Resulting Problem

Existing frameworks for verifying quantum circuits, such as [11–19] generally assume that the circuit contains only unitary operations. Ultimately, only then is it possible to characterize the functionality of a quantum circuit as a unitary matrix. With the availability of dynamic circuit primitives for conducting quantum computations, the question arises as to how circuits using these primitives can be verified. After all, resets, measurements, and classically controlled operations are all non-unitary operations. As such, existing techniques cannot be applied in an out-of-the-box fashion.

Several theoretical works on quantum program and protocol verification exist that deal with dynamic quantum circuits, e.g., [20, 21]. However, their goal is to prove the correctness of an algorithm, i.e., proving that it "works", rather than to check the equivalence of two circuits. Recent work on the equivalence of dynamic quantum circuits based on quantum Mealy machines [22] and ensembles of linear operators [23] shows promise, but has only been evaluated on toy examples ($\approx$10 qubits) and has not led to available software packages for equivalence checking yet. In this chapter, we show that reinventing the wheel is not necessary to allow using existing techniques and tools in combination with dynamic circuits. To this end, we propose two different schemes targeting two slightly different verification scenarios.

First, we consider the question whether two circuits G and G' that might contain non-unitaries are functionally equivalent as a whole—an important question when, e.g., evaluating alternative realizations of certain building blocks in large quantum algorithms. Here, it must be ensured that the alternative realization has the exact same functionality given *any* input. As already shown in Chap. 13, given two circuits G and G' that contain only unitary operations, this reduces to a comparison between the corresponding unitary matrices U and U'. We will show in Sect. 18.2 that any circuit containing non-unitary operations can be transformed to a circuit only containing unitary operations and no intermediate measurements by combining well-known results from quantum information theory. In this way, all existing techniques for verifying the equivalence of two (static) quantum circuits are kept applicable for the broader class of dynamic circuits.

While the above technique conceptually allows to verify circuits containing non-unitaries, it requires to extend a circuit's description by as many qubits as it contains mid-circuit resets. Due to the exponential scaling of the resulting unitary functionality, the complexity of verifying such instances may prove too much to handle for existing tools. The following observation helps to derive an alternative for these cases: In most quantum algorithms, the initial state of the computation can be assumed to be a fixed state (e.g., $|0\ldots0\rangle$). Hence, it might not be necessary at all to ensure that two circuits are *fully* functionally equivalent, but rather that they produce the same distribution of measurement outcomes for the fixed input state, i.e., that they behave the same when executed on a quantum computer. In Sect. 18.3, we show that the probability distribution of a circuit containing non-unitaries can be iteratively extracted from classically simulating the circuit using any available classical quantum circuit simulator.

18.2   Unitary Reconstruction via Circuit Transformation

Dynamic circuit primitives allow one to reuse qubits over the course of a quantum computation and to influence the execution based on classical measurement outcomes. To employ existing verification tools to verify circuits using these primitives, the circuit descriptions G and G' must be transformed to facilitate comparisons of the form $U =^? U'$. This is accomplished by transforming the dynamic circuit primitives to unveil the underlying unitary functionality.

Reset operations pose the first hurdle to overcome. Algorithmically, a reset can be interpreted as measuring a qubit, applying an X operation conditioned on the measurement result being $|1\rangle$ and then discarding the measurement result. Theoretically, any reset operation can be replaced by introducing a new qubit and applying all subsequent operations involving the qubit to be reset to the new qubit. In this fashion, any n-qubit circuit containing r reset instructions can be transformed to a circuit acting on $n + r$ qubits containing no reset primitives.

Example 18.3 Consider again the circuit for the IQPE algorithm from Example 18.2. By iteratively replacing each of the reset operations with new qubits and translating all subsequent gates to the newly introduced qubits, the following circuit acting on four qubits results:

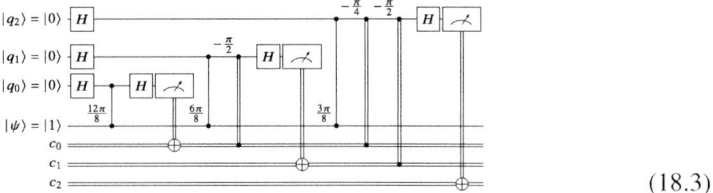

$$(18.3)$$

Once qubit reuse is eliminated from a dynamic circuit, the only potentially non-unitary primitives remaining are mid-circuit measurements and classically controlled operations conditioned on their result. To eliminate these operations, we resort to one of the most fundamental results in quantum computing: the *deferred measurement principle* [4]. This principle states that delaying measurements until the end of a quantum computation does not affect the probability distribution of outcomes. As a consequence, it follows that measurement and classical conditioning on its result commute. Thus, any mid-circuit measurement can be delayed until the very end of the quantum circuit—replacing any classically controlled operations along the way by proper quantum operations controlled by the respective qubit.

Example 18.4 Assume that all reset operations of the IQPE circuit from Example 18.2 have been eliminated, e.g., by transforming the circuit as described in Example 18.3. Then, applying the deferred measurement principle in order to delay all measurements to the end of the circuit and replacing the phase rotations controlled by the measurement outcomes with phase gates controlled on the respective qubits results in the circuit shown below, which is free of dynamic circuit primitives.

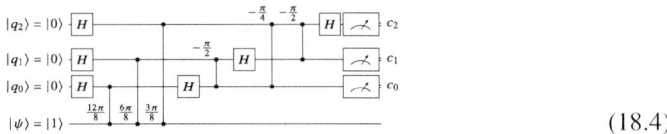

$$(18.4)$$

By combining both aforementioned steps, i.e., substituting reset operations with "fresh" qubits and applying the deferred measurement principle, any dynamic quantum circuit (including non-unitaries) can be transformed to a representation composed of unitary descriptions only. For one, this allows one to verify that a dynamic circuit actually realizes the intended functionality of its static counterpart.

Example 18.5 Compare the transformed circuit obtained in Example 18.4 with the original QPE algorithm shown in Example 18.1. Because they are actually the same, it is easy to conclude that both circuits are indeed equivalent.

Note that it might seem that the proposed approach merely reverses the circuit construction or compilation process. As witnessed in Example 18.5, there is a one-to-one relation between the transformed version of the IQPE and the original QPE circuit. As such, it could be argued that there is nothing to gain from using the technique. However, this is not the case, as almost no assumptions are made about the relation between G and G' in general. In fact, the only requirement is that the transformed versions of both circuits have the same number of primary inputs and outputs. The proposed transformation scheme "touches" nothing but reset, measurement, and classically controlled operations—which are "reversed".

Conceptually, this approach allows to verify circuits containing non-unitaries, at the cost of extending a circuit's description by as many qubits as it contains mid-circuit resets. Since the resulting unitary functionality scales exponentially with the number of qubits, the complexity of verifying such instances increases rapidly. However, this is an inevitable increase whenever verifying whether a dynamic implementation (acting on n_{dyn} qubits and using r resets) still realizes the same functionally as a static counterpart (acting on n_{static} qubits). Since in that case, $n_{dyn} + r = n_{static}$, the proposed scheme augments the dynamic circuit just enough to facilitate comparisons of the form $U \stackrel{?}{=} U'$.

18.3 Extracting Measurement Outcome Distributions via Simulation

Although verification methodologies such as [11–13] frequently allow one to reduce the complexity of verification by exploiting the reversibility of quantum operations, they may not be able to handle this immense complexity in the worst case. Motivated by the fact that most high-level quantum algorithms assume a fixed input state, we argue that it might be sufficient to show that two realizations of such an algorithm produce the same distribution of measurement probabilities given the fixed input state. Verifying that two circuits G and G', which only contain unitary operations, produce equivalent probability distributions given a particular input state $|\psi\rangle$ amounts to classically simulating both computations with $|\psi\rangle$ as input and computing the overlap between the measurement probabilities described by the resulting state vectors.

However, in the presence of dynamic circuit primitives, the concept of a state vector responsible for producing the circuit's measurement outcome distribution (e.g., the probabilities of the individual bitstrings in the IQPE algorithm) does no longer make sense. This is due to the non-unitary nature of the dynamic circuit primitives that no longer allow deterministically simulating the quantum circuit in one go using quantum circuit simulators such as [24–26]. For example, each time a reset operation is encountered, this would technically require the calculation of the partial trace of the system over the particular qubit (and reinitializing it to $|0\rangle$). However, the partial trace is an operation that maps pure states to mixed states. One possible approach for solving this problem would be to repeatedly simulate the dynamic circuit and stochastically realize dynamic circuit primitives such as measurements and resets. However, one would have to perform huge amounts of individual runs in order to reason about the output distribution in a statistically significant way. Another approach requires leaving the pure state picture and using a density matrix simulator (such as, e.g., [27–29]). Although these simulators can naturally handle resets, mid-circuit measurements, and classic-controlled operations, they also do not allow determining the complete distribution of (intermediate) measurement outcomes via a single simulation run, but only the density matrix for a particular set of measurements.

In the following, we propose a technique that allows one to extract the complete set of measurement probabilities for a dynamic circuit given a particular input state. To this end, consider a quantum circuit G involving m measurements. Then each measurement during the circuit simulation constitutes a branching point where the probabilities of the qubit to be measured are check-pointed and the simulation is split into two independent simulations: one assuming that the measurement outcome is $|0\rangle$ and the other assuming the outcome is $|1\rangle$. Depending on the outcome being $|0\rangle$ or $|1\rangle$, a subsequent reset operation is translated into a no-op or an X gate, while any classically controlled operation is ignored or applied, respectively. The probability of observing a particular basis state $|i\rangle = |(i_{m-1} \dots i_0)_2\rangle$ can then be reconstructed from the product of the check-pointed probabilities along the simulation path corresponding to the outcomes i_0 to i_{m-1}.

Example 18.6 Consider again the IQPE algorithm for estimating the phase θ of $U = p(\frac{3\pi}{8})$ corresponding to the eigenstate $|\psi\rangle = |1\rangle$ up to a precision of three bits, as shown in Example 18.2. The circuit contains a total of $m = 3$ measurements (necessary for the 3-bit precision) and uses the fixed input state $|000\rangle \otimes |\psi\rangle = |0001\rangle$. Iteratively simulating the circuit, check-pointing the probabilities at each of the measurements, and adjusting the subsequent circuit parts to be simulated accordingly, results in the computational flow:

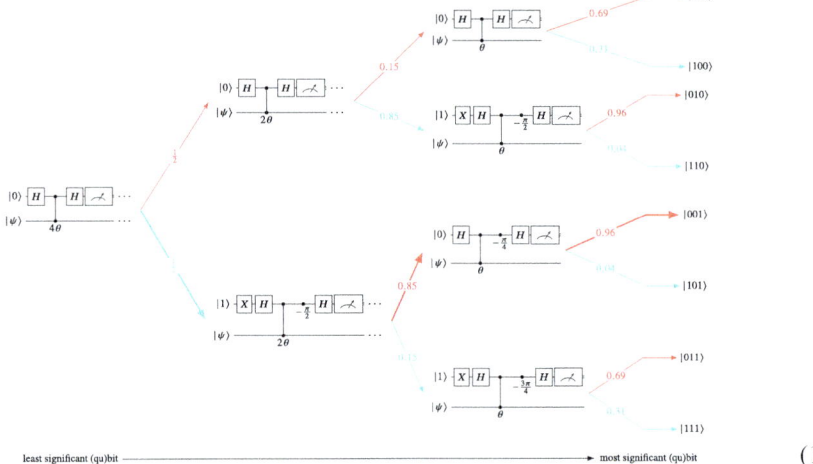

$$(18.5)$$

Here, red arrows denote the $|0\rangle$-successor, while blue arrows denote the $|1\rangle$-successor, i.e., the subsequent computations upon measuring $|0\rangle$ or $|1\rangle$, respectively. The path indicated in bold represents the extraction of the probability for the $|001\rangle$ state—resulting in $\frac{1}{2} * 0.85 * 0.96 \approx 0.408$.

Extracting the distribution of measurement outcomes of a dynamic circuit in this fashion naturally requires a total of 2^m individual simulations, where m is the number of mid-circuit measurements. However, large parts of the simulations can be shared between simulation runs. For example, the circuit up to the first checkpoint only needs to be simulated once, while two simulations are necessary up to the second checkpoint, and so on. In general, the kth sub-circuit needs to be simulated in at most 2^k variations. If any measurement along a path produces a probability of zero, further simulations along that path need not be started at all. In addition, the individual simulations in between checkpoints are completely independent of one another, and hence are embarrassingly parallelizable. On top of that, each of these sub-circuits consists of a much smaller number of gates and acts on far fewer qubits than the whole dynamic circuit's static counterpart. As a consequence, the complete measurement outcome distribution can be efficiently extracted in many cases, although exponentially many simulations might be required in the worst case.

18.4 Summary of Results

The methods proposed above have been implemented on top of the open-source quantum circuit equivalence checking tool QCEC [30] that was introduced in Chap. 14. In [3], experimental evaluations have been carried out to demonstrate the feasibility of the proposed methods. To this end, various instances of the famous Bernstein-Vazirani algorithm [31], the Quantum Fourier Transform, and the QPE algorithm, which was used as a running example throughout this chapter, have been considered as benchmarks. For each static algorithm, a dynamic realization has been derived [9, 32, 33]. These algorithms are a good fit for evaluating the overhead of the proposed schemes, as they feature all the hurdles of dynamic quantum circuits that have to be overcome for verifying their equivalence. The following conclusions can be drawn from these evaluations:

- The resulting evaluations demonstrate that transforming the dynamic circuit using the proposed scheme incurs practically no overhead (the runtime for the transformation is on the order of 1 ms for all considered instances) and allows one to successfully verify the full functional equivalence of the Bernstein-Vazirani and QFT algorithms with up to 128 qubits in a fraction of a second. Even QPE instances with up to 50 qubits can be verified in <3 min.

- The results for the Bernstein-Vazirani and QPE algorithms show that extracting the complete measurement probabilities of a dynamic circuit can, in fact, be faster than classically simulating the static counterpart by more than an order of magnitude. This is in line with the discussions at the end of Sect. 18.3 and can be attributed to the fact that the resulting state vectors are sparse, i.e., feature only few nonzero amplitudes. In contrast, the state vector resulting from the QFT is dense, which immediately reflects in the runtime of the extraction scheme, i.e., it roughly doubles with every added qubit. Thus, the scheme of Sect. 18.2 should be preferred in this case.

Overall, the proposed schemes form a generic solution for handling non-unitaries when verifying the equivalence of quantum circuits.

Implementation, Usage, Documentation, and Results

 The proposed schemes for handling non-unitaries are available as part of the open-source MQT QCEC tool [30] at
https://github . com/munich-quantum-toolkit/qcec,
which can be installed using pip install mqt . qcec.

 The resulting tool can be setup as described and illustrated in section 14.5 on page 117. Compared to the example shown in section 14.5, only a single line has to be modified to transform dynamic quantum circuits as proposed in this chapter, i.e., instead of only calling

```
qcec.verify(qc1, qc2)
```

the verification function can be configured with

```
qcec.verify(qc1, qc2, transform_dynamic_circuit
    =True)
```

to transform any dynamic circuit before running the verification.

 Documentation on all available configuration options is available at
https://mqt . readthedocs . io/projects/qcec

 Details on the experimental setup, evaluations, and results can be found in [3].

References

1. A.W. Cross, A. Javadi-Abhari, T. Alexander et al., OpenQASM 3: a broader and deeper quantum assembly language (2021) arXiv preprint: arXiv:2104.14722
2. IBM Quantum, Quantum circuits get a dynamic upgrade with the help of concurrent classical computation (2021). https://www.ibm.com/blogs/research/2021/02/quantum-phase-estimation/
3. L. Burgholzer, R. Wille, Handling non-unitaries in quantum circuit equivalence checking, in *Design Automation Conference* (2022). https://doi.org/10.1145/3489517.3530482
4. M.A. Nielsen, I.L. Chuang, *Quantum Computation and Quantum Information* (Cambridge University Press, 2010)
5. P.W. Shor, Polynomial-time algorithms for prime factorization and discrete logarithms on a quantum computer. SIAM J. Comput. (1997). https://doi.org/10.1137/S0097539795293172
6. A.W. Harrow, A. Hassidim, S. Lloyd, Quantum algorithm for linear systems of equations. Phys. Rev. Lett. **103**(15) (2009)
7. S. Lloyd, M. Mohseni, P. Rebentrost, Quantum principal component analysis. Nat. Phys. **10**(9), 631–633 (2014)
8. C.H. Bennett, G. Brassard, C. Crépeau, R. Jozsa, A. Peres, W.K. Wootters, Teleporting an unknown quantum state via dual classical and Einstein-Podolsky-Rosen channels. Phys. Rev. Lett. **70**(13) (1993)
9. M. Dobsicek, G. Johansson, V.S. Shumeiko, G. Wendin, Arbitrary accuracy iterative phase estimation algorithm as a two qubit benchmark. Phys. Rev. A (2007)
10. A.D. Corcoles, M. Takita, K. Inoue et al., Exploiting dynamic quantum circuits in a quantum algorithm with superconducting qubits (2021). arXiv preprint: arXiv:2102.01682
11. S. Yamashita, I.L. Markov, Fast equivalence-checking for quantum circuits, in *International Symposium on Nanoscale Architectures* (2010). https://doi.org/10.1109/NANOARCH.2010.5510932
12. L. Burgholzer, R. Wille, Advanced equivalence checking for quantum circuits, in *IEEE Transaction on CAD of Integrated Circuits and Systems* (2021). https://doi.org/10.1109/TCAD.2020.3032630
13. L. Burgholzer, R. Raymond, R. Wille, "Verifying results of the IBM Qiskit quantum circuit compilation flow, in *International Conference on Quantum Computing and Engineering* (2020). https://doi.org/10.1109/QCE49297.2020.00051
14. G.F. Viamontes, I.L. Markov, J.P. Hayes, Checking equivalence of quantum circuits and states, in *International Conference on CAD* (2007)
15. P. Niemann, R. Wille, R. Drechsler, Equivalence checking in multi-level quantum systems, in *International Conference of Reversible Computation* (2014)
16. S.-A. Wang, C.-Y. Lu, I.-M. Tsai, S.-Y. Kuo, An XQDD-based verification method for quantum circuits, in *IEICE Transactions on Fundamentals* (2008), pp. 584–594. https://doi.org/10.1093/ietfec/e91-a.2.584
17. K.N. Smith, M.A. Thornton, A quantum computational compiler and design tool for technology-specific targets,' in *International Symposium on Computer Architecture* (2019), pp. 579–588
18. M. Amy, Towards large-scale functional verification of universal quantum circuits, in *International Conference on Quantum Physics and Logic* (2019)
19. X. Hong, X. Zhou, S. Li, Y. Feng, M. Ying, A tensor network based decision diagram for representation of quantum circuits (2020). arXiv preprint: arXiv:2009.02618
20. M. Ying, Floyd–Hoare logic for quantum programs. ACM Trans. Program. Lang. Syst. (2011)
21. M. Ying (ed.), *Foundations of Quantum Programming* (2016)
22. Q. Wang, R. Li, M. Ying, Equivalence checking of sequential quantum circuits (2021). arXiv preprint: arXiv:1811.07722

23. X. Hong, Y. Feng, S. Li, M. Ying, Equivalence checking of dynamic quantum circuits (2021). arXiv preprint: arXiv:2106.01658
24. A. Zulehner, R. Wille, Advanced simulation of quantum computations, in *IEEE Transactions on CAD of Integrated Circuits and Systems* (2019). https://doi.org/10.1109/TCAD.2018.2834427
25. G.G. Guerreschi, J. Hogaboam, F. Baruffa, N.P.D. Sawaya, Intel quantum simulator: a cloud-ready high-performance simulator of quantum circuits. Quantum Sci. Technol. **5**(3), 034 007 (2020), ISSN: 2058-9565, https://doi.org/10.1088/2058-9565/ab8505
26. B. Villalonga, S. Boixo, B. Nelson et al., A flexible high-performance simulator for verifying and benchmarking quantum circuits implemented on real hardware. npj Quantum Inform. (2019), ISSN: 2056-6387, https://doi.org/10.1038/s41534-019-0196-1
27. G.F. Viamontes, I.L. Markov, J.P. Hayes, Graph-based simulation of quantum computation in the density matrix representation, in *Quantum Information and Computation, Part II* (2004)
28. T. Grurl, J. Fuß, R. Wille, Considering decoherence errors in the simulation of quantum circuits using decision diagrams, in *International Conference on CAD* (2020)
29. A. Li, O. Subasi, X. Yang, S. Krishnamoorthy, Density matrix quantum circuit simulation via the BSP machine on modern GPU clusters, in *International Conference for High Performance Computing, Networking, Storage and Analysis* (2020)
30. L. Burgholzer, R. Wille, QCEC: a JKQ tool for quantum circuit equivalence checking. Software Impacts (2021)
31. E. Bernstein, U. Vazirani, Quantum complexity theory. SIAM J. Comput. (1997)
32. P. Nation, B. Johnson, How to measure and reset a qubit in the middle of a circuit execution. IBM Research Blog. (2021), https://www.ibm.com/blogs/research/2021/02/quantum-mid-circuit-measurement/
33. R.B. Griffiths, C.-S. Niu, Semiclassical Fourier transform for quantum computation. Phys. Rev. Lett. (1996)

Variational Quantum Algorithms [1] have been proposed to achieve a computational advantage despite the high gate error rates and short coherence times of today's NISQ [2] devices. These algorithms use quantum programs as subroutines in a classical optimization routine to solve problems in chemistry [3], finance [4], discrete optimization [5], and more. Before running such an algorithm on a target device, a compilation step has to be performed—a costly and non-trivial procedure (cf. Part III). To avoid recompiling a quantum circuit in each iteration step of a variational algorithm, the circuit is usually compiled once in the *parameterized* form in which the parameters tuned by the classical optimization routine are not bound to specific values. This compiled parameterized circuit can then be used in the optimization loop without having to perform all compilation steps over and over again.

The increasing use of parameterized circuits in the development of quantum algorithms has also brought about the need to verify that these circuits have been compiled correctly. Established equivalence checking methods such as those proposed in [6–10] and in the previous chapters are capable of proving that a compiled circuit remains within the specification after compilation. However, these approaches cannot handle parameterized circuits. The only way to verify the compilation of variational quantum algorithms with these methods is to check the equivalence of the original and the instantiated compiled ansatz in each iteration. Since equivalence checking is a difficult problem even for one instance, solving it repeatedly is hardly a feasible approach.

In this chapter (based on [11]), we propose, for the first time, a methodology to solve the equivalence checking problem for parameterized quantum circuits. The proposed multi-stage method starts out by using an approach based on the ZX-calculus [12, 13] to try and prove the equivalence of both parameterized circuits. If this does not succeed, the circuit parameters are instantiated, and a conventional complete equivalence checker is employed. To make this check as easy as possible, we derive an instantiation scheme that allows one to simplify most of the parameterized gates from the circuits. Furthermore, we prove that this yields a complete equivalence checking procedure for parameterized quantum circuits.

The remainder of this chapter is structured as follows: Sect. 19.1 provides some background on variational quantum algorithms. Then, Sect. 19.2 motivates the problem of checking the equivalence of parameterized quantum circuits, while Sect. 19.3 proposes a complete equivalence checking flow for parameterized circuits. After elaborating on the proposed methods in Sect. 19.4, Sect. 19.5 concludes with details on the implementation and discussion on experimental results.

19.1 Variational Quantum Algorithms

Quantum computations in variational algorithms are expressed via *ansatz circuits*. These are shallow *parameterized quantum circuits* $G(\theta)$ with a parameter vector $\theta = (\theta_0, \dots, \theta_{p-1})$. Each *assignment* $\sigma : \{\theta_i \mid 0 \le i < p\} \to (-\pi, \pi]^p$ yields an *instantiated circuit* $G(\sigma(\theta))$. The goal of a variational algorithm is to successively adapt the circuit parameters using a classical optimization routine (e.g., gradient descent) so that the resulting circuit can be used to estimate some desired quantity, e.g., the ground state of a molecule. Using classical computing, variational ansatz circuits can be deliberately kept shallow. However, shallow circuits are generally less expressive than more complex circuits, and ansatz circuits of different complexity have been developed.

Example 19.1 Consider the following parameterized quantum circuit:

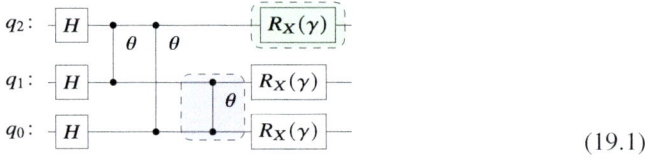

$$\tag{19.1}$$

It represents an instance of an *Quantum Alternating Operator Ansatz* (QAOA, Hadfield et al. [5]) which can be used to solve problems such as quadratic unconstrained binary optimization problems. The name comes from the fact that a QAOA ansatz is made up of two circuit blocks, each of which is parameterized by a different parameter.

Now assume that this circuit shall be compiled to a linear five-qubit architecture that supports a gate library consisting of CNOT, R_Z, and H gates. Then, one possible compiled circuit is given by:

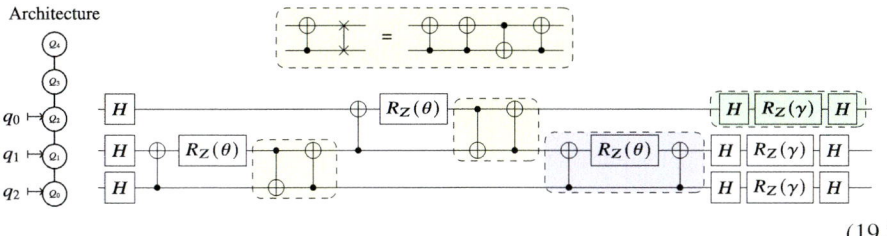

$$(19.2)$$

Because the ZZ gate (which is synthesized to two CNOT gates and an R_Z gate) between physical qubit Q_2 and Q_0 cannot be executed directly, a SWAP gate has to be introduced into the circuit. As shown above the circuit, this SWAP gate is synthesized as a sequence of three CNOT gates. One of these CNOTs can be canceled with the CNOT of the synthesized ZZ gate, simplifying the circuit in the process.

Note that, throughout the iterations in the variational algorithm, only the values of the parameters change, while the general circuit structure of the ansatz stays the same. Therefore, it is common to perform the costly compilation of an ansatz in its parametric form only once and then instantiate the parameters of this compiled circuit in each iteration instead of instantiating the parameters and compiling each time. This saves a lot of overhead incurred by repeated and potentially expensive compilation steps.

19.2 Verifying Variational Quantum Circuits

Compiling a variational ansatz can significantly change its structure due to synthesized gates, SWAP insertions, and optimizations applied during the compilation process. Because compilation errors are hard to detect from the results of an iteration of a variational algorithm alone, it is paramount to ensure a priori that the compiled ansatz still adheres to its specification.

However, when parameterized gates are allowed in the circuits to be checked, the equivalence checking problem becomes even more general. Checking the equivalence of two parameterized circuits $G(\theta)$ and $G'(\theta)$ requires showing that

$$G(\sigma(\theta_0), \ldots, \sigma(\theta_{p-1})) \text{ and } G'(\sigma(\theta)) \qquad (19.3)$$

are equivalent for *all* assignments σ. The naive approach to circumventing this problem would be to construct the matrices of $G(\theta)$ and $G'(\theta)$ symbolically. But then, in addition to the exponential size of the matrices, one also has to deal with symbolic variables when constructing the matrices. As is known from computer algebra systems, trying to represent symbolic matrix entries precisely requires a lot of space for storing the coefficients of the symbolic variables.

Alternatively, one might be tempted to simply check the equivalence for one specific instantiation of two parameterized circuits to conclude the equivalence. Unfortunately, this brings about a couple of difficult challenges in itself. On the one hand, instantiating parameters in a *random* or *unstructured* fashion produces circuits that are hard—if not impossible—to check with existing methods. On the other hand, instantiating parameters non-randomly can, as the following example shows, lead to false positives and, hence, also does not provide a sufficient solution.

Example 19.2 Consider the following incorrect application of a commutation rule for the R_Z gate:

$$R_Z(\beta) \oplus R_Z(\alpha) \oplus \quad \rightarrow \quad R_Z(\alpha) \oplus R_Z(\beta) \oplus \tag{19.4}$$

Even though the two circuits are not equivalent for all $\alpha, \beta \in (-\pi, \pi]$, they are equivalent if $\alpha = \beta$. Therefore, the equivalence of parametric circuits cannot be decided by checking the equivalence of any instantiation.

Because variational ansatz circuits are instantiated with different parameters in each iteration, verifying the compilation results of such circuits without symbolic equivalence checking methods requires checking the equivalence of $G(\sigma_i(\theta_0), \ldots, \sigma_i(\theta_{p-1}))$ and $G'(\sigma_i(\theta))$ for the parameter assignment σ_i in each iteration i of the hybrid optimization loop. Repeatedly checking equivalence in this fashion leads to obvious problems—in particular when checking a single instance is already costly. Therefore, dedicated methods for equivalence checking of variational ansatz circuits are desperately needed.

19.3 An Equivalence Checking Method for Parameterized Quantum Circuits

In this section, a fully automated, efficient, and *complete* method for checking the equivalence of arbitrary parameterized quantum circuits is proposed. To this end, a dedicated multi-stage process, as shown in Fig. 19.1, is used. The main ideas behind this method are sketched below.

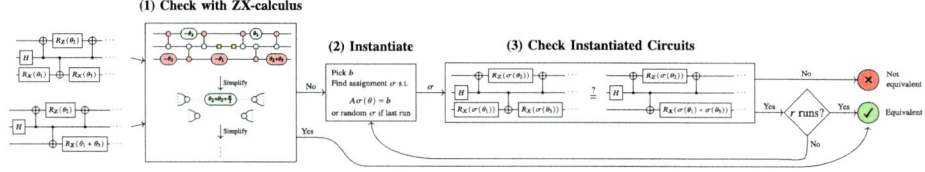

Fig. 19.1 Proposed equivalence checking method for parameterized quantum circuits

First, parameterized circuits are converted into parameterized ZX-diagrams—a graphic description of quantum processes—that are then checked with a parametric version of the ZX-calculus simplification algorithm described in [9, 14] (step (1) in Fig. 19.1). If this check yields an affirmative answer to the question of equivalence, then the two parameterized circuits have been proven to be equivalent, and the algorithm terminates. However, due to the incompleteness of this rewriting approach, a non-affirmative answer does, in general, not imply that the circuits in question are non-equivalent. Therefore, if the ZX-calculus approach cannot prove equivalence, further steps must be taken. To this end, note that, to prove the non-equivalence of two parameterized circuits $G(\theta)$ and $G'(\theta)$ it suffices to find *one* assignment $\sigma : \theta \to (-\pi, \pi]^n$ such that $G(\sigma(\theta)) \neq G'(\sigma(\theta))$.

As discussed above, choosing a *good* assignment is a delicate issue, as random assignments may produce hard equivalence checking instances, and non-random assignments can (as discussed in Example 19.2) lead to false positives. However, instantiation can never produce false negatives. Hence, rather than choosing a random assignment, one can take advantage of the degrees of freedom provided by the parameters and instantiate the circuits in such a fashion as to make the subsequent equivalence check as simple as possible. In the proposed method, this is achieved by solving a linear system obtained from the expressions in parameterized gates (step (2) in Fig. 19.1). The instantiated circuits are then checked with an existing, complete equivalence checking method (step (3) in Fig. 19.1). If this method manages to prove non-equivalence of the instantiated circuits, it can be concluded that the parameterized circuits are not equivalent either.

To counteract the possibility of false positives, multiple non-random instantiations are checked. In the worst case, however, all of these instantiations could yield false positives. Therefore, the circuits are instantiated randomly as the last resort before being checked one last time. This has the disadvantage of handing very complicated circuits over to the equivalence checker, but the advantage—as proven later in this chapter—is that the probability of obtaining a false positive through random instantiation is zero. Hence, the output of this last check is returned as the final result.

19.4 Implementation and Completeness

Having the ideas and concepts discussed above (and illustrated in Fig. 19.1), this section now provides details of the implementation of the respective steps and proves that the overall methodology resulting is indeed complete.

19.4.1 Equivalence Checking of Parameterized Circuits with the ZX-Calculus

This section describes how equivalence of parameterized circuits can be checked with the ZX-calculus.

Example 19.3 The circuit shown in Example 19.1 can be translated into a ZX-diagram and simplified by applying the spider fusion rule.

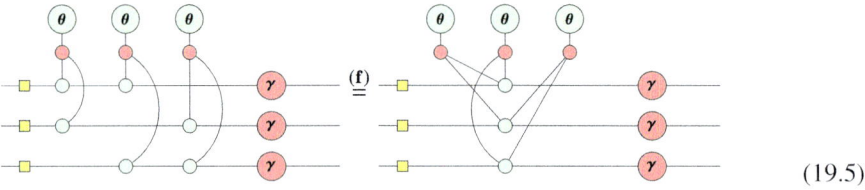

$$\tag{19.5}$$

Although this ZX-diagram looks very similar to the original circuit, the ZZ gates have been translated into a form (so-called phase gadgets) that has no direct interpretation as a quantum circuit anymore.

As shown previously in Chap. 17, given two quantum circuits G and G', one can check them for equivalence by constructing the ZX-diagram of $G^{-1}G'$ and, then, simplifying the diagram as much as possible using the rewriting system shown in Fig. 17.1. If the ZX-diagram can be reduced to the identity diagram (the ZX-diagram consisting only of wires and no spiders), the two circuits are shown to be equivalent. This method can be used to handle parameterized circuits as well, since all phases $\alpha_i, \beta_i, \gamma_i$ in Fig. 17.1 are just symbolic. For the sake of finding matches for the rewriting rules their precise value does not matter.

Whether or not the ZX-diagram of the circuit $G^{-1}(\theta)G'(\theta)$ can be rewritten to the identity diagram depends on the symbolic expressions appearing in the diagram. It is reasonable to assume that all phases α_k are linear functions of the parameters $\theta = (\theta_0, \dots, \theta_{p-1})$, i.e., they are of the form

$$\alpha_k = \left(\sum_{i=0}^{p-1} c_i \theta_i \right) + d \qquad c_0, \dots, c_{p-1}, d \in \mathbb{R}. \tag{19.6}$$

This assumption is justified considering that many optimizations of quantum circuits involve gate commutations and summation of rotation angles for rotation gates [15]. Under this assumption it is easy to see that all phases resulting from the application of the rules shown in Fig. 17.1 are also just linear functions of these parameters. If some parameters cancel and the resulting phase has the form $k\frac{\pi}{2}, k \in \mathbb{Z}$, new simplifications become possible. If, eventually, all parameters cancel and the diagram is reduces to the identity when simplifying

$G^{-1}(\theta)G'(\theta)$, it can be concluded that $G(\theta)$ is equivalent to $G'(\theta)$ entirely in parameterized form.

19.4.2 Determining Instantiation Parameters

Established equivalence checking methods like [8] greatly benefit from circuits that have a certain repeating structure in their functionality [14, 16]. When using such methods to check equivalence of instantiated circuits, one can exploit the degrees of freedom in the parameters to try to create such structures. As shown in experimental evaluations (which are summarized later in Sect. 19.5), this can significantly decrease the complexity of checking the resulting parameter-free instance.

Based on previous assumptions, if the rotation angles are of the form $\alpha_k = \left(\sum_{i=0}^{p-1} c_i \theta_i \right) + d_k$ for some constant d_k, then one can try to instantiate the angles α_k to a predefined value by solving a linear system. This system has as many equations as there are parameterized angles in the circuits. When choosing the angles to solve for, one has to distinguish two cases in checking non-equivalent circuits:

- The error in the circuit appears in one of the parameter-free gates. In this case, one can try to solve for $\alpha_k = 0$, for all parameterized angles. This effectively removes the entire parameterized gates—leading to a simpler circuit.
- The error in the circuit appears in one of the parameterized gates. In this case, removing the gates from the circuit would mask the error and lead to false conclusions. Then, one can still try to solve for rotation angles of 0. Nevertheless, when solving the linear system, some of the equations will, in practice, not have a solution precisely because of erroneous optimizations. The gates that cannot be removed will, therefore, usually be the ones containing an error.

In general, the linear system can be overdetermined and may not have a solution. It is NP-hard in general to find a solution such that the maximal number of equations are satisfied [17]. One can still try to satisfy equations in a greedy fashion which, although not optimal, can be done in polynomial time.

19.4.3 Completeness Proof

If both the ZX-checker and the previously described instantiation method fail to show that two parameterized circuits are not equivalent, then it cannot be concluded with absolute certainty that the circuits are equivalent. However, in the following, we prove that it suffices to randomly instantiate the parameters and check the equivalence of the resulting parameter-free circuits. The core observation is that, statistically, two random instantiations can be proven to *almost never* produce a subsequent false positive. Here, "almost never" means that out of the infinitely many choices for the real-valued parameters θ_i, the probability of choosing a combination of values that results in a false positive is zero.

The following assumes a familiarity with complex analysis and measure theory to keep the proofs as concise as possible. Any reader unfamiliar with these concepts might skip to the last paragraph of this section.

Lemma 19.1 *Let $f : \mathbb{C}^n \to \mathbb{C}$ be an analytic function, λ_n be the Lebesgue measure on $\mathbb{R}^n$, and $Z(f) = \{x \in \mathbb{R}^n \mid f(x) = 0\}$ the set of real zeros of f. If $\lambda_n(Z(f)) > 0$ then, $f = 0$.*

Proof Since $\lambda_n(Z(f)) > 0$ implies that $Z(f)$ contains an accumulation point, this lemma follows directly from the identity principle for analytic functions. □

The converse of this lemma states that a non-trivial analytic function can have only countably many real zeros. With this lemma, we can now show the desired result.

Theorem 19.1 *Let $G(\theta)$ and $G'(\theta)$ be two non-equivalent quantum circuits with parameter vector $\theta \in (-\pi, \pi]^n$. Suppose that all rotation angles in the gates of $G(\theta)$ and $G'(\theta)$ are linear functions of θ. Then, $\mathbb{P}\{\theta \mid G(\theta) = G'(\theta)\} = 0$.*

Proof Due to the assumption on the angles of $G(\theta)$ and $G'(\theta)$, all matrix entries of their respective gate matrices are of the form $e^{i \sum_{i=0}^{n} c_i \theta_i + d}$, which is a complex analytic function in θ. The system matrices $U(\theta)$ and $U'(\theta)$ are the product of the respective gate matrices of the circuits G and G'. Therefore, all entries of U and U' are analytic in θ. Without loss of generality, consider the i, jth entries $u_{i,j}(\theta)$, $u'_{i,j}(\theta)$. Then, $u_{i,j}(\theta) - u'_{i,j}(\theta)$ is also analytic. By Lemma 19.1, the set of zeros of $u_{i,j}(\theta) - u'_{i,j}(\theta)$ has measure zero. Therefore, the probability of choosing random parameters in $(-\pi, \pi]^n$ such that $u_{i,j}(\theta) = u'_{i,j}(\theta)$ is 0. Since this holds for one matrix entry, it follows immediately that $\mathbb{P}\{\theta \mid G(\theta) = G'(\theta)\} = 0$. □

The attentive reader might argue that this fact negates the need for the previously discussed methods. While this is true *in theory*, our experimental evaluations clearly demonstrate that, in practice, equivalence checking of circuits with random rotations is a computationally difficult task. It is, therefore, only used as a last resort in the proposed equivalence checking method.

19.5 Summary of Results

The equivalence checking method proposed in this chapter has been implemented on top of the open-source MQT QCEC tool [18] introduced in Chap. 14. To test a wide range of parameterized ansatz circuits, the proposed method was evaluated on the *entire* available library of parameterized ansatz circuits provided by the Qiskit circuit library [19]. These evaluations have shown:

- The proposed approach scales to the largest quantum systems available today and is capable of verifying the compilation results of any of the variational algorithms that are currently being explored for near-term applications. All considered circuits were successfully verified completely with the ZX-calculus, and no instantiation was necessary.
- While the method proposed in Chap. 14 used to conduct the equivalence check of the instantiated circuits quickly runs into the set timeout of 1 h when using random instantiation, the instantiation approach through solving linear systems allows to conclude the non-equivalence for a much larger range of circuits—in many cases within fractions of a second. This underlines the point made at the end of the previous section that, while random instantiations would be "good enough" for equivalence checking parameterized circuits in theory, they are hardly practical.

Overall, the proposed method yields a complete equivalence checking scheme by combining a ZX-calculus approach working directly on parameterized circuits and an instantiation strategy to create parameter-free circuits that are efficiently checkable by existing equivalence checking methods. While this method has been developed with variational algorithms in mind, the approach is much more general and works for any application that uses parameterized quantum circuits.

Implementation, Usage, Documentation, and Results

 The resulting methodology for verifying the equivalence of parametrized quantum circuits is available as part of the open-source MQT QCEC tool [18] at https://github . com/munich-quantum-toolkit/qcec, which can be installed using pip install mqt . qcec.

 The resulting tool can be setup as described and illustrated in section 14.5 on page 117. Support for parametrized quantum circuits is enabled by default. Configuring the number of instantiations only requires the modification of a single line: Instead of only calling

```
qcec . verify ( qc1 ,  qc2 )
```

the verification function can be configured with

```
qcec . verify ( qc1 ,  qc2 ,  additional_instantiations =3)
```

to perform a maximum of three additional instantiations after trying to set all parameters to zero and before resorting to a completely random instantiation of the parameters.

 Documentation on all available configuration options is available at https : //mqt . readthedocs . io/projects/qcec

 Details on the experimental setup, evaluations, and results can be found in [11].

References

1. M. Cerezo, A. Arrasmith, R. Babbush et al., Variational quantum algorithms (2020). arXiv preprint: arXiv:2012.09265
2. J. Preskill, Quantum computing in the NISQ era and beyond. Quantum **2**, 79 (2018)
3. S. McArdle, S. Endo, A. Aspuru-Guzik, S.C. Benjamin, X. Yuan, Quantum computational chemistry. Rev. Mod. Phys. **92**(1), 015 003 (2020). https://doi.org/10.1103/RevModPhys.92.015003
4. D. Egger, C. Gambella, J. Marecek et al., Quantum computing for finance: state-of-the-art and future prospects. IEEE Trans. Quantum Eng. (2020). https://doi.org/10.1109/TQE.2020.3030314
5. S. Hadfield, Z. Wang, B. O'Gorman, E.G. Rieffel, D. Venturelli, R. Biswas, From the quantum approximate optimization algorithm to a quantum alternating operator Ansatz. Algorithms **12**(2), 34 (2019). ISBN: 1999-4893, https://doi.org/10.3390/a12020034
6. L. Burgholzer, R. Raymond, R. Wille, Verifying results of the IBM Qiskit quantum circuit compilation flow, in *International Conference on Quantum Computing and Engineering* (2020). https://doi.org/10.1109/QCE49297.2020.00051
7. M. Amy, Towards large-scale functional verification of universal quantum circuits, in *International Conference on Quantum Physics and Logic* (2019)
8. L. Burgholzer, R. Wille, Advanced equivalence checking for quantum circuits, in *IEEE Transaction on CAD of Integrated Circuits and Systems* (2021). https://doi.org/10.1109/TCAD.2020.3032630
9. A. Kissinger, J. van de Wetering, Reducing T-count with the ZX-calculus. Phys. Rev. A (2020)
10. W. Chun-Yu, T. Yuan-Hung, J. Chaio-Shan, J. Jie-Hong, Accurate BDD-based unitary manipulation for scalable and robust quantum circuit verification, in *Design Automation Conference* (2022). https://doi.org/10.1145/3489517.3530481
11. T. Peham, L. Burgholzer, R. Wille, Equivalence checking of parameterized quantum circuits: verifying the compilation of variational quantum algorithms, in *Asia and South Pacific Design Automation Conference* (2023). https://doi.org/10.1145/3566097.3567932
12. J. van de Wetering, ZX-calculus for the working quantum computer scientist (2020). arXiv preprint: arXiv:2012.13966
13. B. Coecke, A. Kissinger, Picturing quantum processes, in *Diagrammatic Representation and Inference*, ed. by P. Chapman, G. Stapleton, A. Moktefi, S. Perez-Kriz, F. Bellucci (2018)
14. T. Peham, L. Burgholzer, R. Wille, Equivalence checking paradigms in quantum circuit design: a case study, in *Design Automation Conference* (2022)
15. Y. Nam, N.J. Ross, Y. Su, A.M. Childs, D. Maslov, Automated optimization of large quantum circuits with continuous parameters. npj Quantum Inf. (2018)
16. L. Burgholzer, R. Raymond, I. Sengupta, R. Wille, Efficient construction of functional representations for quantum algorithms, in *International Conference of Reversible Computation* (2021)
17. V. Guruswami, P. Raghavendra, Hardness of solving sparse overdetermined linear systems: a 3-Query PCP over integers, in *Electronic Colloquium on Computational Complexity* (2009), https://eccc.weizmann.ac.il/report/2009/020
18. L. Burgholzer, R. Wille, CEC: a JKQ tool for quantum circuit equivalence checking. Software Impacts (2021)
19. A. Javadi-Abhari, M. Treinish, K. Krsulich et al., Quantum computing with Qiskit (2024). arXiv:2405.08810 [quant-ph]

This part of the book explores the challenge of verifying the correctness of quantum circuits, which requires software solutions that can efficiently and automatically check the equivalence of two quantum circuits. At first glance, the characteristics of quantum computing, such as the use of superposition and entanglement, make verification more challenging than in classical computing. At the same time, the fact that quantum operations are reversible offers a potential that is not available in classical computing. This potential can be used for efficient equivalence checking. More precisely, the following contributions were presented:

- A general equivalence checking methodology has been proposed that explicitly takes advantage of quantum characteristics to efficiently determine whether two quantum circuits are equivalent or not (cf. Chap. 14).
- Based on that, a fine-tuned strategy has been proposed to verify the results of the compilation of quantum circuits from information on how a given circuit G is compiled to a resulting implementation G' (cf. Chap. 15). The resulting method was demonstrated to verify instances with more than ten thousand gates within seconds—even if optimizations are employed that are not directly accounted for. In contrast to the classical realm, this could eventually make verifying the results of sophisticated design flows feasible in general.
- It was demonstrated, theoretically and empirically, that many errors in quantum circuits can already be detected by a few simulations with randomly chosen initial states. Three quantum stimuli generation schemes have been proposed that offer a trade-off between error detection rate (as well as the required number of stimuli) and efficiency (cf. Chap. 16).

- Additionally, the proposed core methodology is further enriched by an approach using the ZX-Calculus as a complementary alternative that is best used in tandem with the other proposed approaches (cf. Chap. 17).
- Finally, the methodology has been extended to support dynamic quantum circuits (cf. Chap. 18) as well as parametrized quantum circuits (cf. Chap. 19)—substantially broadening the applicability of verification for those important classes of circuits.

All the efforts listed above have contributed to the development of the open-source quantum circuit equivalence checking tool QCEC [1], which is available at https://github.com/ munich-quantum-toolkit/qcec. QCEC constitutes a push-button, accessible, state-of-the-art verification tool for quantum circuits. Hopefully, the development of these kinds of software tools will prevent quantum computing from developing a verification gap similar to that of classical systems. https://github.com/munich-quantum-toolkit/qcec

Reference

1. L. Burgholzer and R. Wille, "QCEC: A JKQ tool for quantum circuit equivalence checking," *Software Impacts* (2021)

Part V
Conclusions

Design automation tools and software have been crucial for the development of classical circuits and systems. They enable faster and more reliable design cycles, reduce human errors, and allow for complex and large-scale designs. In the domain of quantum computing, the corresponding design automation methods (which have been developed over the past decades) remain heavily underutilized. The *Munich Quantum Toolkit* (MQT) made substantial contributions towards leveraging this latent potential. This book provided a detailed look into some of the inner workings of this toolkit.

For three key design tasks (namely, classical simulation, compilation, and verification of quantum circuits), several methods and tools have been presented that explicitly use design automation expertise, such as decision diagrams, SAT encodings, as well as reasoning engines. More precisely:

For classical simulation of quantum circuits, the book explored the use of decision diagrams, a dedicated data structure inspired by classical design automation, to accelerate the simulation of quantum circuits on classical machines. This task is challenging because the quantum state and operations are commonly represented by vectors and matrices that have an exponential size with respect to the number of qubits. Therefore, efficient methods and powerful resources are required to simulate quantum circuits classically. In the book, several approaches have been proposed that significantly improve the current state of the art in classical quantum circuit simulation and further establish decision diagrams as a core data structure for this task.

For the task of compilation, a novel approach has been proposed, which determines optimal solutions to one of the most important and challenging problems in this process: mapping quantum circuits to architectures that have limited connectivity between their qubits. By encoding this NP-complete problem in a symbolic fashion and by using powerful

L. Burgholzer and R. Wille, *Design Automation Tools and Software for Quantum Computing*, https://doi.org/10.1007/978-3-032-06770-8_21

reasoning engines to cope with the vast complexity of the underlying search space, the resulting solution allows one to establish lower bounds on the achievable performance and to evaluate the quality of the results from existing and future heuristic methods. Since this solution is hardly scalable due to the immense search space, the book further demonstrated how to limit the search space of the optimal mapping problem to drastically improve the runtime of the resulting method while, at the same time, preserving optimality.

Finally, design automation methods for the verification of quantum circuits have been investigated. It is essential for the successful future of quantum computing to ensure the correctness of quantum circuits, e.g., throughout the whole compilation flow. By that, the emerge of a verification gap for quantum circuits, i.e., a situation where the physical development of a technology substantially outperforms our ability to design suitable applications for it or to verify it, can hopefully be avoided or substantially mitigated. To this end, the book has proposed several complementary approaches that form the first comprehensive suite of efficient methods and automated tools for quantum circuit verification.

Overall, this book showcased the inner workings of the MQT and demonstrated the immense benefits of leveraging existing knowledge and expertise in classical circuit and system design for the development of quantum software. All implementations described in this book (and many more) are available as open-source solutions and can be accessed at github.com/munich-quantum-toolkit.

Batch number: 10176978

Printed by Printforce, the Netherlands